Investigating School Mathematics

ROBERT E. EICHOLZ

PHARES G. O'DAFFER

CHARLES R. FLEENOR

ADDISON-WESLEY PUBLISHING COMPANY

MENLO PARK, CALIFORNIA • READING, MASSACHUSETTS • LONDON • DON MILLS, ONTARIO

Contents

Let's Explore your Mathematics Book 2

1 Counting and Measurement
Counting to find length 4
Using a ruler 8
Measuring to nearest unit and half unit 12
Counting squares to find area 16
Fractions in measurement 18
Counting cubes to find volume 24
Liquid measure: cup to gallon 26
Solving Story Problems 27
Reviewing the Ideas 28

2 Place Value
Grouping by tens 30
Two-, three-, and four-digit numerals 32
Inequalities: one- to four-digit numerals 42
Using large numbers 44
Solving Story Problems 46
Reviewing the Ideas 48
Keeping in Touch 49

3 Addition and Subtraction
Inverse relation between addition and subtraction 50
Differences and missing addends 56
Basic principles for addition 60
Rearranging addends 62
The Function Machine 70
Reviewing the Ideas 72
Keeping in Touch 73

4 Geometry
Edges, faces, and vertices 74
Simple geometric figures 76
Segments 78
Angles 80
Triangles 82
Reviewing the Ideas 88
Keeping in Touch 89

5 Adding and Subtracting
Dimes and pennies — tens and ones 90
Sums and differences 94
Keeping in Touch 96
Addition with regrouping 98
Regrouping to find differences 106
Solving Story Problems 108
Money problems 112
Reviewing the Ideas 118
Keeping in Touch 120

6 Multiplication
Multiplication and equivalent sets 122
The number line 124
Multiplication and intersections of strips 126
Repeated addition 128
Factors and products 132
Keeping in Touch 135
Zero and one principles for multiplication 136
Commutative (order) principle for multiplication 138
Rearranging factors 140
Multiplication-addition principle 142
Multiplication combinations (for products through 81) 146
Short picture problems 156
Patterns and multiplication facts 158
Product sets 162
Solving Story Problems 164
Reviewing the Ideas 166
Keeping in Touch 168

7 Division
Division and equivalent sets 170
Rectangular arrays and division 172

Copyright © 1973 by Addison-Wesley Publishing Company, Inc. Philippines Copyright 1973
All rights reserved. No part of this publication may be reproduced, stored in a retrieval system, or transmitted, in any form or by any means, electronic, mechanical, photocopying, recording, or otherwise, without the prior written permission of the publisher.
Printed in the United States of America. Published simultaneously in Canada.

BCDEFGHIJKL76543

Division and repeated subtraction	174
Division on the number line	176
Quotients as missing factors	178
Inverse relation between multiplication and division	180
Keeping in Touch	182
Division of a set into equivalent sets	186
Solving Story Problems	190
Reviewing the Ideas	192
Keeping in Touch	194

8 Geometry

Parallel lines	196
Angles and parallel lines	198
Quadrilaterals	200
Special quadrilaterals	202
Parallelograms	204
Polygons	206
Simple closed curves	208
Symmetry	210
Reviewing the Ideas	212
Keeping in Touch	213

9 Number Theory

Odd and even numbers	214
Multiples	218
Factors	220
Prime numbers	222
Reviewing the Ideas	224
Keeping in Touch	225

10 Multiplying

Multiplying by 10 and 100	226
The multiplication-addition principle	234
Multiplication algorithm	240
Two-, three-, and four-digit factors	248
Estimation	250
Reviewing the Ideas	252
Keeping in Touch	254

11 Geometry and Graphing

Using coordinates	256
Graphing number pairs	260
Point pictures	262
Symmetry	264
Translations	266
Graphing functions	268
Bar graphs	270
Negative numbers and graphing	272
Reviewing the Ideas	274
Keeping in Touch	275

12 Dividing

Division concepts	276
Inverse relation between multiplication and division	278
Input, output, and rule	280
Keeping in Touch	282
Division and repeated subtraction	284
Finding quotients	286
Solving Story Problems	290
Keeping in Touch	292
Developing the long-division algorithm	294
Solving Story Problems	298
Division with remainders	300
Checking division	302
Reviewing the Ideas	304
Keeping in Touch	306

Mathematical Activities 309

Appendix

More Practice	A-1
Books to Explore	A-38
Glossary	A-41
Tables of Measures	A-44
Index	A-45

Illustrations by Sue Gilmour, Reid Fancher, and Patti Dwyer. Photographs by Marshall Berman. Grateful acknowledgment is made to the Palo Alto Unified School District for allowing the photographing of pupils at Addison Elementary School.

● *Let's explore your mathematics book.*

Investigating the Ideas

This is a sample lesson. It will help you understand how you will use your book.

In this part of a lesson there are things for you to **investigate** and discover.

 Can you find some Investigations where you would use these objects?

Scissors

Colored strips

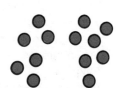
Counters

Discussing the Ideas

In this part of a lesson you will **discuss the ideas** you investigated. You will be sharing your ideas with others. You are getting ready to **use the ideas**.

1. Look through your book. Can you explain an easy way to find the "Investigating the Ideas" sections?

2. Can you explain an easy way to find the "Discussing the Ideas" sections in your book?

3. Find a "Discussing the Ideas" section that begins at the top of a page.

4. Find a page called "Keeping in Touch." What do you think this means?

Using the Ideas

In this part of the lesson you will **use the ideas**. You will work problems to improve your understanding of the ideas you have discussed. Try these.

1. How many "Investigating the Ideas" sections are in Chapter 4?

2. Find the number of "Discussing the Ideas" sections in Chapter 6.

3. How many "Using the Ideas" sections are in Chapter 8?

4. Look up *prime numbers* in your index. What page numbers are given?

Problems in these boxes are a **special challenge** for you. Be sure to try some of them. See if you can do this one.

think

The faces of a block are numbered from 1 to 6. What is the sum of the numbers you cannot see?

1 Counting and Measurement

● *Can you measure by counting?*

Investigating the Ideas

Choose one of your colored strips.

Count how many times you have to put your strip down to move it across your desk top.

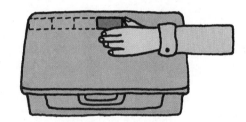

 Can you use one of your strips to measure some other parts of your desk?

Discussing the Ideas

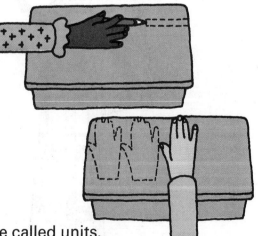

1. How did you use your strip to measure your desk?

2. Jane said, "My desk is about 4 pencils wide." What did Jane mean?

3. How many hands wide is your desk?

4. Objects used for measuring are called units. What are the three units used so far in this lesson?

4

Using the Ideas

1. Count the number of strip units for the measure of each object.

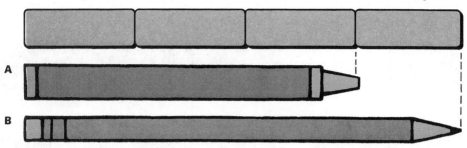

2. How many strip units long is each object?

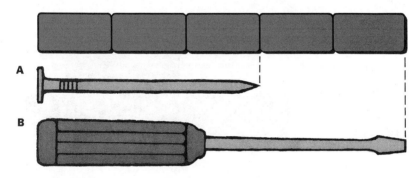

3. How many paper clip units long is the straw?

4. Count the ⊢—⊣ units to measure each rod.

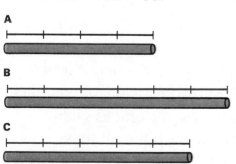

think

If the measure of the green rod is 6 units, guess the measure of the yellow rod.

● How can you make and use your own ruler?

Investigating the Ideas

Each of these rulers uses a different strip as unit.

RED STRIP

PURPLE STRIP

Cut a long strip of paper. Choose one of these strips and make a ruler as shown.

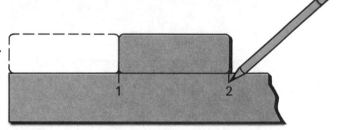

Making a Light Green Ruler

 Can you use your ruler to find the length of your pencil?

Discussing the Ideas

1. Which unit would give the largest number for the length of your pencil?
2. Which unit would give the smallest number for the length of your pencil?
3. Joe measured his crayon using a red ruler. He found it was 4 units long. What is the length of his crayon using a purple ruler?
4. Bill and Jane each made a ruler. Jane used a unit as wide as her hand. Bill used a unit as wide as his finger. If they both measure the same thing, who will get the larger number?

Using the Ideas

1. Draw five lines on your paper as shown. Number your lines 1 to 5.

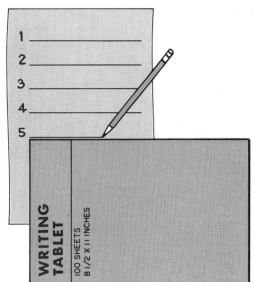

2. On the top line use your ruler to mark 2 **points** that are 4 units apart. Name your points *A* and *B*.

You have a **segment** *AB* that is 4 units long.

3. A On line 2 mark a segment *CD* that is 3 units long.

   B On line 3 mark a segment *EF* that is 5 units long.

   C On line 4 mark a segment *GH* that is more than 3 units and less than 4 units long.

   D On line 5 mark a segment that is more than 2 units and less than 3 units long.

4. Try this one without using any of the rulers you made.

   A How long is the rope if the red strip is the unit?

   B How long is the rope if the light green strip is the unit?

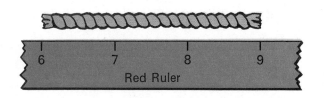

● *Let's think about some special units.*

Investigating the Ideas

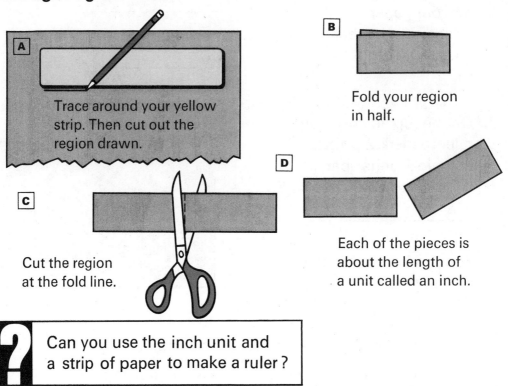

A Trace around your yellow strip. Then cut out the region drawn.

B Fold your region in half.

C Cut the region at the fold line.

D Each of the pieces is about the length of a unit called an inch.

? Can you use the inch unit and a strip of paper to make a ruler?

Discussing the Ideas

1. How does the ruler you made compare to an inch ruler?

2. Can you name some objects that are usually measured in
   A inches?   B feet?   C yards?

3. A mile is 1760 yards. A train of 120 boxcars is about 1 mile long. What are some distances that are usually measured in miles?

4. Which unit would you choose to measure
   A a football field?   C the width of your room?
   B a pencil?   D the distance from New York to California?

Using the Ideas

1. Find the length in inches of each object.

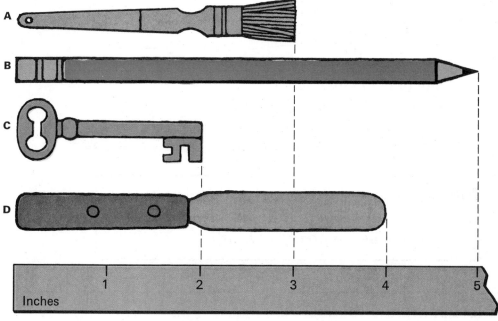

2. Use your inch ruler to measure each segment.

   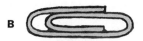

★ 3. If this ⊢——⊣ is your unit, find the measure of each object.

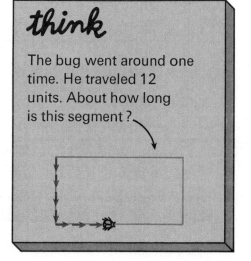

think

The bug went around one time. He traveled 12 units. About how long is this segment?

● *How does the centimeter compare to the inch?*

Investigating the Ideas

This is a centimeter unit. ⟶ ⊢—⊣

This is an inch unit. ⟶ ⊢———⊣

Find one of your strips that is
just as long as the centimeter unit.

> **?** Can you use that strip or the centimeter unit to make a centimeter ruler and then use it to find the length of this brush?

Discussing the Ideas

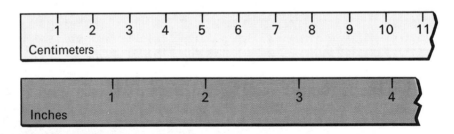

1. Which is longer, the inch or the centimeter?

2. Which ruler will give the greater number for the length of the brush?

3. The crayon is about 3 inches long. Measure it with a centimeter ruler.

Did you get more or less than 3?

Using the Ideas

1. Use your inch ruler to measure each object.

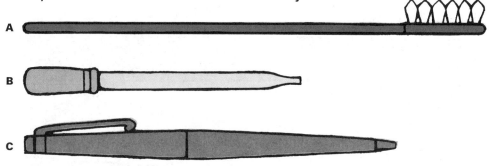

2. Use your centimeter ruler to measure each object.

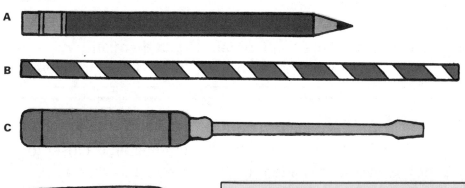

★ G

What comes next?
1. AA, AB, AC, AD, AE, _?_
2. AR, BS, CT, DU, EV, _?_
3. AZ, BY, CX, DW, EV, _?_
4. AA, AB, BB, BC, CC, _?_
5. AB, DE, GH, JK, MN, _?_
6. AB, DC, EF, HG, IJ, _?_

● *How can you measure to the nearest unit?*

Investigating the Ideas

Half-inch ($\frac{1}{2}$ inch) marks on your ruler help you find measures to the nearest inch.

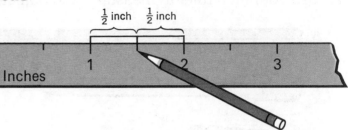

Put half-inch marks on your ruler. Use it to tell whether the clothespin is closer to 2 or to 3 inches.

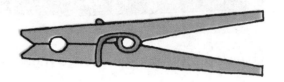

 Can you measure some other objects to the nearest inch? | List them and record their measures.

Discussing the Ideas

1. Explain how you could tell that the clothespin is closer to 3 inches. We say: The length to the nearest inch is 3.

2. The length of the spring is **more than** 1 inch but **less than** 2 inches. Is it closer to 1 or to 2?

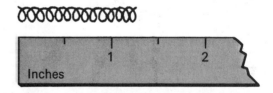

3. The length of the spring (to the nearest inch) is ▨ inches.

4. Is the nail closer to 4 or to 5 inches?

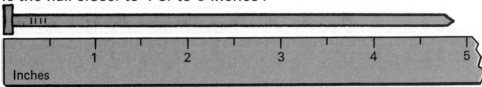

5. The length of the nail (to the nearest inch) is ▨ inches.

Using the Ideas

1. Give the length of each object to the nearest inch.

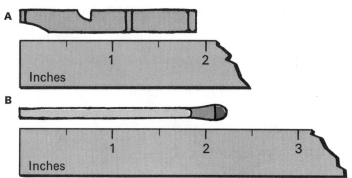

2. Measure each segment to the nearest inch.

    A _____   B _____
    C _____
    D _____
    E _____   F _____   G _____

3. Give the measure of each object to the nearest centimeter.

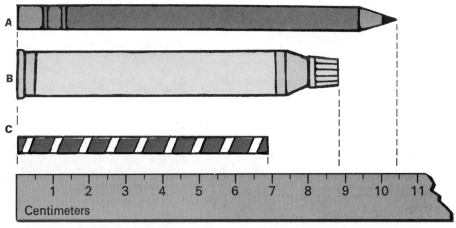

4. Put half-centimeter marks on your centimeter ruler. Use it to measure these objects to the nearest centimeter.

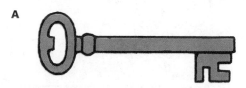   

More practice, page A-1, Set 1

● *How can you measure to the nearest half unit?*

Investigating the Ideas

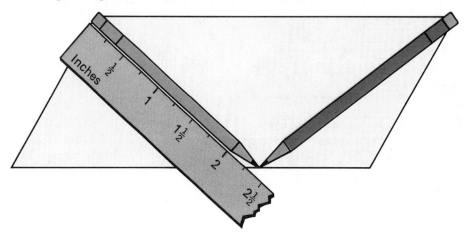

 Look at the **blue** pencil and the ruler. Which pencil do you think is longer? Use your ruler to check your guess.

Discussing the Ideas

1. A Is the end of the needle closer to 1 or to $1\frac{1}{2}$? (Read $1\frac{1}{2}$ as "one and one half.")
   B What is the length of the needle **to the nearest half inch**?

2. A Is the end of the toothpick closer to $1\frac{1}{2}$ or to 2?
   B What is the length of the toothpick **to the nearest half inch**?

3. A Does the figure show that the width of the nickel is nearer to $\frac{1}{2}$ (one-half) inch or to 1 inch?
   B What mark on the ruler helps you to decide your answer?

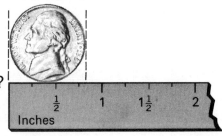

Using the Ideas

1. Use your inch ruler to measure each object to the nearest half inch.

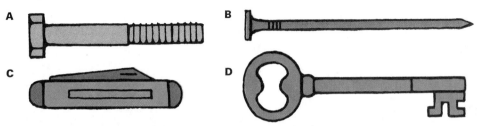

2. Use your inch and centimeter rulers to help you answer these questions.
   - A How long is 4 centimeters to the nearest half inch?
   - B About how many inches long is 5 centimeters?
   - C How many inches long is 9 centimeters to the nearest half inch?

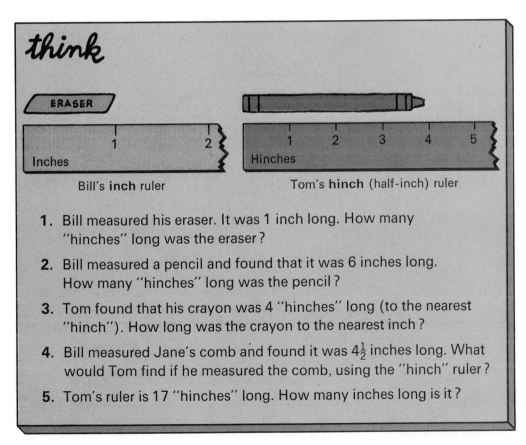

think

Bill's **inch** ruler

Tom's **hinch** (half-inch) ruler

1. Bill measured his eraser. It was 1 inch long. How many "hinches" long was the eraser?

2. Bill measured a pencil and found that it was 6 inches long. How many "hinches" long was the pencil?

3. Tom found that his crayon was 4 "hinches" long (to the nearest "hinch"). How long was the crayon to the nearest inch?

4. Bill measured Jane's comb and found it was $4\frac{1}{2}$ inches long. What would Tom find if he measured the comb, using the "hinch" ruler?

5. Tom's ruler is 17 "hinches" long. How many inches long is it?

More practice, page A-2, Set 2

● *What does it mean to find area?*

Investigating the Ideas

Trace this region and cut it into four 1-inch squares.

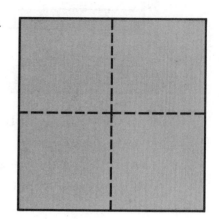

 How many different-shaped regions can you make using the 4 squares? Draw a picture of each one you find.

Discussing the Ideas

1. We find the **area** of a region by counting **square units**.

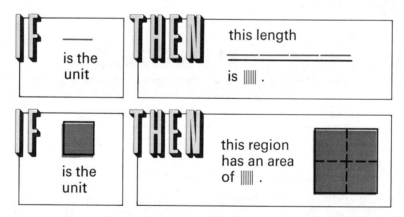

2. What is the area of each region you found in the Investigation?

3. If each small square is a unit, find each area. Explain how you found it.

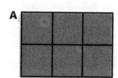
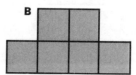
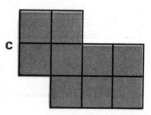

16

Using the Ideas

1. The unit in these exercises is the square centimeter. Find the area of each shaded region.

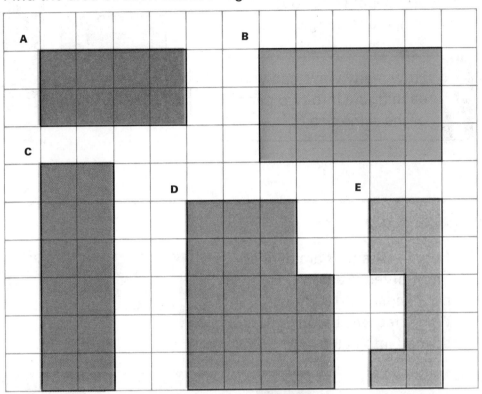

2. Draw a region that has an area of 6. Use the same unit as in exercise 1.

★ 3. If the area of this region is 1, ⟶

what is the area of this region? ⟶

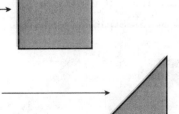

17

● *Let's explore using fractions in finding area.*

Investigating the Ideas

Use graph paper to cut out a square and a rectangle.

How many ways can you fold each figure into two parts of the same size?

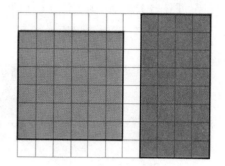

Discussing the Ideas

1. Regions *A* and *B* are divided into **halves**. Region *C* is not divided into **halves**. Which regions below are divided into halves?

   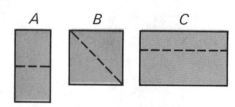

2. If this region has an area of 4, what is the area of this region?

3. Explain how to find the area of each shaded region. Each small square is 1 unit.

   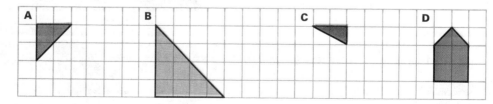

Using the Ideas

1. The units for these exercises are marked with gray lines. Find the area of each shaded region.

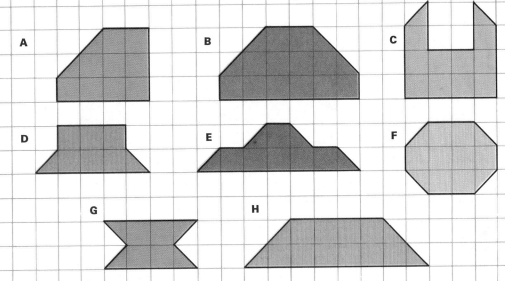

2. Find the area of each region.

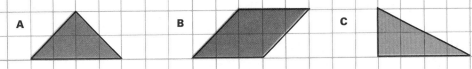

★ **3.** Find the area of each region.

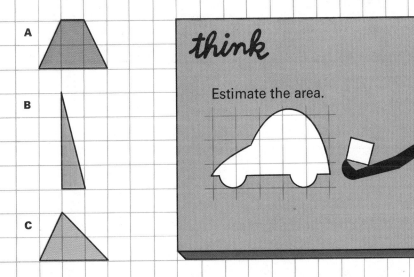

think

Estimate the area.

More practice, page A-2, Set 3

● *How can you use halves and fourths in measurement?*

Investigating the Ideas

Trace each region on paper and then cut it out.

  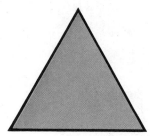

 Can you fold each region into 4 parts that are the same size and shape?

Discussing the Ideas

1. The inch ruler has been divided into **fourths**.

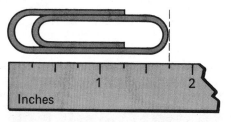 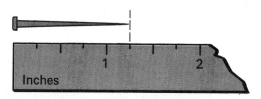

The length of the paper clip is $1\frac{3}{4}$ inches. (Read $1\frac{3}{4}$ as "one and three fourths.")

What is the length of the pin?

2. A Find the area.

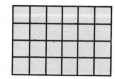

B If you fold the rectangle into fourths, what is the area of each fourth?

C What is the area of $\frac{3}{4}$ of the region?

Using the Ideas

1. Find each length to the nearest one-fourth inch.

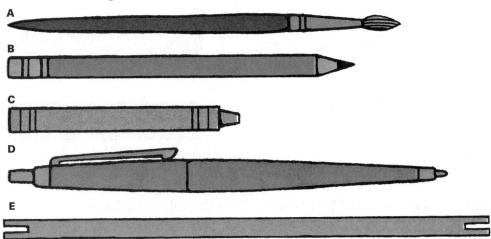

2. A Find the area of this square.
   B What is the area of $\frac{1}{2}$ of it?
   C What is the area of $\frac{1}{4}$ of it?

3. A Find the area of this rectangle.
   B What is the area of $\frac{1}{2}$ of it?
   C What is the area of $\frac{1}{4}$ of it?
   ★ D What is the area of $\frac{3}{4}$ of it?

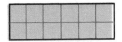

4. A What is the area of this square?

   ★ B What is the area of $\frac{1}{2}$ of it?
   ★ C What is the area of $\frac{1}{4}$ of it?

The area of this region is 9. Draw a picture of the unit.

● *Let's find out more about fractions.*

Investigating the Ideas

Give the missing fraction.

1. If you fold like this, each part is ▥ of the paper.

2. If you fold like this, each part is ▥ of the paper.

 Cut out a strip of paper. Can you fold it so that each part is $\frac{1}{8}$ (one eighth) of the paper?

Discussing the Ideas

1. Explain what you would do if you wanted to color $\frac{3}{8}$ (three eighths) of the paper you folded.

2. A Give the fraction that tells what part of this strip is colored.

   B What part of the strip is not colored?

3. What part of each region is shaded?

Using the Ideas

1. Give the fraction that tells what part of each region is colored.

A B C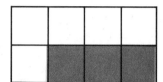

D E 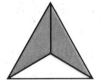 F

2. Jim folded a rectangle into eighths. He colored part of it and then cut it into squares like these.

   A What fraction tells the part of the rectangle he colored?

   B What part of the set of squares is not colored?

3. Give the fraction that tells what part of this set of squares is colored.

Short Stories

1 Had 6 baseball cards. Gave away $\frac{1}{2}$ of them. How many left?

2 Store had 6 lollipops. Bought $\frac{1}{3}$ of them. Bought how many?

3 8 children. $\frac{1}{4}$ of them wear glasses. How many wear glasses?

4 12 cookies. $\frac{1}{4}$ of them are chocolate. How many are chocolate?

More practice, page A-3, Set 4

● What does it mean to find volume?

Investigating the Ideas

How many blocks does it take to make each figure?

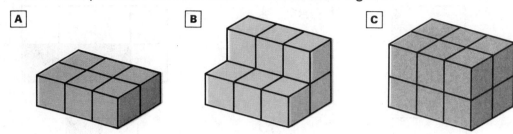

 How many different-shaped figures can you make using four blocks?

Discussing the Ideas

1. Give the missing numbers. Explain your answers.

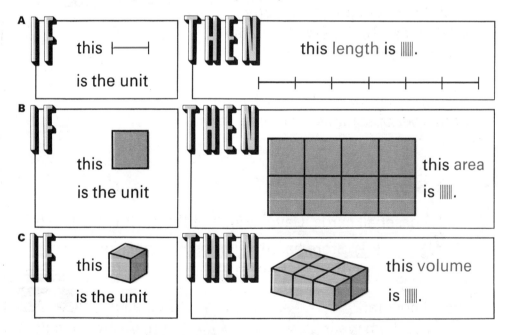

2. What is the volume of each figure in the Investigation?

Using the Ideas

The unit used in these exercises is . Find the number of cubic units (volume) in each figure.

1.

2.

3.

4.

5.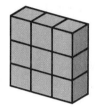

6.

7.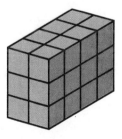

8.

9.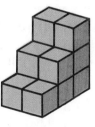

★ 10.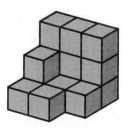

★ 11.

● *What are some other units for measuring volume?*

Investigating the Ideas

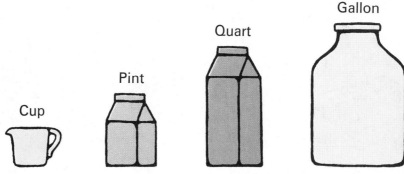

Use these four different-sized containers.

 Can you work with your classmates and fill out a measurement table like this one?

Liquid Measurement

1 pint holds __?__ cups.
1 quart holds __?__ pints.
1 gallon holds __?__ quarts.

Discussing the Ideas

Use your measurement table to help you answer these questions.

1. A How many cups in a pint?
   B How many pints in a quart?
   C How many cups in a quart?

2. A cup holds 8 ounces.
   A How many ounces in a pint?
   B How many ounces in ½ pint?

3. A How many quarts are in a gallon?
   B Explain how to find how many pints are in a gallon.
   C Explain how to find how many cups a gallon will hold.

Using the Ideas

Shopping Problems

Sally's mother asked her to go to the store.
Here is her shopping list: →

Shopping List
Tomato Juice (large can)
Bread (2 loaves)
Milk (1 gallon)
Eggs (1 carton)
Orange Soda
Grape Drink (1 gallon)

1. Sally picked up 3 quarts of milk. How much more should she get?

2. Sally knew that the orange soda was for her little brother's birthday party. If each of the 6 boys at the party drinks 2 cups of orange soda, how many quart bottles should Sally buy?

3. Sally bought a carton of 6 pint bottles. How many quarts did she buy?

4. Which should Sally buy to save money — the gallon or the 2 half gallons?

★ 5. One can held 28 ounces of tomato juice. Another held 1 quart of tomato juice. Which can held more juice? How much more?

27

Reviewing the Ideas

1. Bob takes steps about this long. ―――――
   Sue's steps are just half as long. ――――→ ―――
   - A Who will take the most steps to cross the room?
   - B Who takes the longer step?
   - C If Bob takes 10 steps to cross the room, how many steps will Sue take?
   - D If Sue takes 12 steps from the door to the teacher's desk, how many steps will Bob take?

2. A Give the length of each segment using the inch as your unit.

   ―――――――――――――――――――――――――

   ――――――――――――――――

   - B Measure each segment to the nearest half centimeter.

   ★ C Ted called his unit the twinch. _____Twinch_____
   Give the measure of each segment above, using the twinch as your unit. Use your ruler if you like.

3. Does the inch or the centimeter give the larger number when you measure the same object?

4. A Is the inch longer than 2 centimeters?
   B Is the inch longer than 3 centimeters?

5. Using the unit shown, give the area for each region.

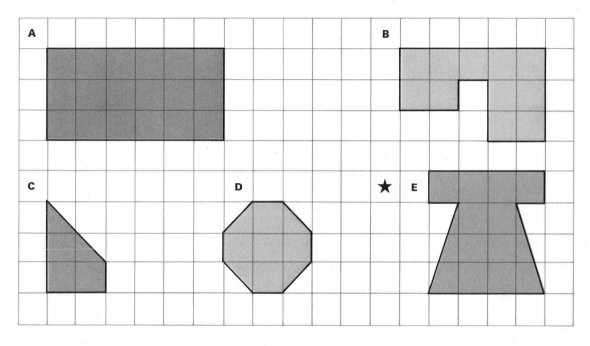

6. Give the volume of each figure below.

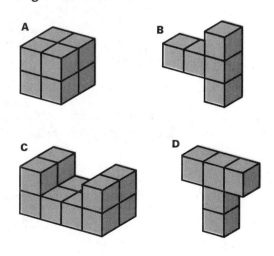

Find the area of this region.

29

2 Place Value

● *Can grouping by tens help you find the number in a set?*

Investigating the Ideas

How good are you at guessing?
1. Guess the number of blue stars quickly.

2. Give the number of red stars quickly.

3. Check your answers. Which was easier?

ten

ten

 Can you draw a set of dots (between 30 and 50) so that a classmate can give the number quickly?

Discussing the Ideas

1. Each set of coins in A, B, and C is worth the same amount.
   A What is the value of each set?
   B Which two sets are easiest to count? Explain your answer.

2. Suppose you had 40 checkers to count. Could you count these by only counting to ten once? Explain.

Using the Ideas

1. Write the numeral for each set of pencils.

   A 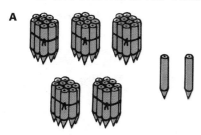   B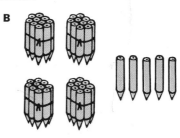

   5 tens and 2   4 tens and 5
   We write ▓.   We write ▓.

2. In the pictures below there are 10 objects in each ring. Give the number of objects in each box.

   A 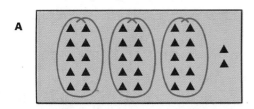   B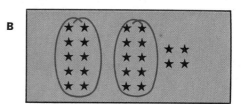

   C 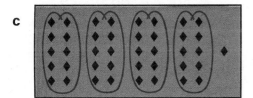   D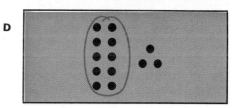

3. Write the 2-digit numeral for each of these.
   - A 2 tens and 3
   - B 4 tens and 2
   - C 3 tens and 4
   - D 6 tens and 7
   - E 9 tens and 3
   - F 5 tens and 7
   - G 1 ten and 1
   - H 1 ten and 0
   - I 2 tens and 0

4. Give the correct digit for each ▓.
   - A 37 means 3 tens and ▓.
   - B 48 means ▓ tens and 8.
   - C 82 means 8 tens and ▓.
   - D 50 means ▓ tens and 0.
   - E 93 means ▓ tens and 3.
   - F 15 means ▓ tens and 5.
   - G 67 means 6 tens and ▓.
   - H 76 means ▓ tens and 6.

More practice, page A-4, Set 5

● *Do you know the number names up to one hundred?*

Discussing the Ideas

1. There are ten sticks in each bundle. Tell the number of tens and then the number of sticks in all.

A	(2 bundles)	E	(5 bundles)
B	(3 bundles)	F	(6 bundles)
C	(4 bundles)	G	(7 bundles)
D	(5 bundles)	H	(8 bundles)

2. Read the number. Then tell how many tens and how many ones.

   A 63 D 57 G 96 J 60 M 17 P 11 S 48 V 35
   B 47 E 21 H 85 K 51 N 42 Q 10 T 80 W 89
   C 39 F 16 I 43 L 30 O 78 R 99 U 27 X 70

3. Give the word name for each of these.

   A 6 tens and 5   E 8 tens and 3   I 7 tens and 0
   B 5 tens and 9   F 3 tens and 0   J 4 tens and 9
   C 4 tens and 2   G 4 tens and 3   K 8 tens and 5
   D 5 tens and 4   H 1 ten and 0   L 7 tens and 3

4. Count by tens to one hundred.
   ten, twenty, thirty, . . .

5. Count by fives to one hundred.
   five, ten, fifteen, twenty, . . .

6. Count by twos to one hundred.
   two, four, six, eight, ten, twelve, . . .

Using the Ideas

1. Find the value of each collection.

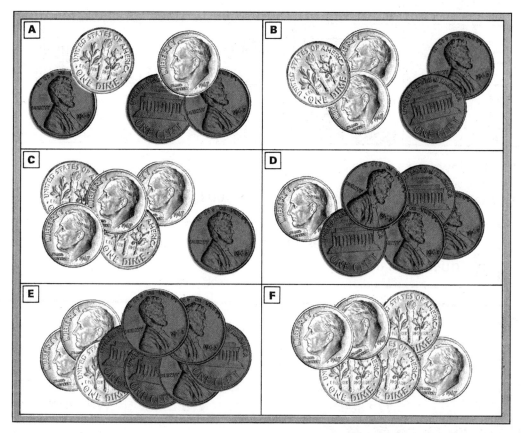

2. Copy each row, giving the missing numbers.

A 4, 5, 6, 7, ▨, ▨, ▨, ▨, 12, 13
B 24, 25, 26, 27, ▨, ▨, ▨, ▨, 32, 33
C 52, 53, 54, 55, 56, 57, ▨, ▨, ▨, ▨
D 90, 91, ▨, ▨, ▨, ▨, 96, 97, 98, 99
E 72, 73, 74, ▨, ▨, ▨, ▨, 79, 80, 81
F ▨, ▨, ▨, ▨, 48, 49, 50, 51, 52, 53
G 38, ▨, ▨, ▨, ▨, 43, 44, 45, 46, 47
H ▨, ▨, ▨, ▨, 62, 63, 64, 65, 66, 67
I ▨, ▨, ▨, ▨, 4, 5, 6, 7, 8, 9
J ▨, ▨, ▨, ▨, 22, 23, 24, 25, 26, 27

think

1. Using the digits 1, 2, 7, 9, how many 2-digit numerals can you write?
2. Which of these numerals names the largest number?
3. Which names the smallest?

More practice, page A-4, Set 6

● *How large is a hundred?*

Investigating the Ideas

How far down the sidewalk will 100 steps take me?

How high up the wall is 100 inches?

How full will the measuring cup be with 100 beans?

How far down the hall is the 100th tile from here?

? How well can you estimate 100? Try one of the questions above.

Discussing the Ideas

1. Give the missing numerals in the table.

We see	We think	We write
(8 bundles)	8 tens	80
(9 bundles)	9 tens	A
(10 bundles)	10 tens	B

2. Explain how to count one hundred pencils without counting higher than ten.

3. How would you draw one hundred dots without counting higher than ten?

4. Can you draw one hundred twenty-three dots without counting higher than ten? Explain.

Using the Ideas

1. Give the missing numerals.
   A For 9 tens and 6, we write ▒▒▒.   D For 9 tens and 9, we write ▒▒▒.
   B For 9 tens and 7, we write ▒▒▒.   E For 9 tens and 10, we write ▒▒▒.
   C For 9 tens and 8, we write ▒▒▒.   F For 10 tens and 0, we write ▒▒▒.

2. Find the total number of cents in each box.

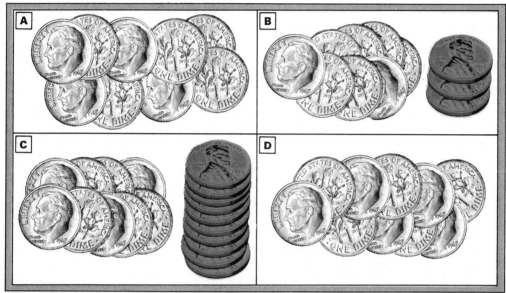

3. Write the missing numerals.
   For 10 tens, we write 100.
   For 20 tens, we write 200.
   A For 30 tens, we write ▒▒▒.
   B For 40 tens, we write ▒▒▒.
   C For 80 tens, we write ▒▒▒.

4. Write the missing numerals.
   For 10 tens, we write 100.
   For 11 tens, we write 110.
   For 12 tens, we write 120.
   A For 13 tens, we write ▒▒▒.
   B For 14 tens, we write ▒▒▒.
   C For 15 tens, we write ▒▒▒.
   D For 16 tens, we write ▒▒▒.

think

The rug covers some of the tiles. How many tiles on this floor?

● *Let's investigate 3-digit numerals.*

Investigating the Ideas

Directions:
Use 3 sets of 9 cards each with the digits 1 through 9. Shuffle the 27 cards and deal 3 cards to each of three players.

Each player forms a 3-digit numeral and guesses whether his number is **High, Middle,** or **Low**.

Then compare numbers and score 1 point for a correct guess. Shuffle and deal again. 10 points wins!

High, Middle, Low Game

High

Middle

Low

? Can you play this game with some classmates?

Discussing the Ideas

1. John and Anne both said their numbers were **High**. Which player won?

John   Anne   Nancy

2. Would Nancy win if she had said **Middle**? Who was middle?

3. Use the figure to answer the questions.
    A How many hundreds?
    B How many tens?
    C How many ones?
    D How many sticks, 254, 542, or 245?

ten
ten
ten
ten

4. Read the number. Then tell how many hundreds, tens, and ones.
   A 278 C 512 E 765 G 318 I 380 K 900 M 707 O 437
   B 346 D 923 F 492 H 640 J 704 L 506 N 770 P 864

Using the Ideas

1. Write the numeral. (*h* stands for hundreds and *t* for tens.)
   - A 3 *h*, 2 *t*, and 6
   - B 4 *h*, 7 *t*, and 2
   - C 6 *h*, 5 *t*, and 0
   - D 6 *h*, 0 *t*, and 1
   - E 3 *h*, 2 *t*, and 0
   - F 3 *h*, 0 *t*, and 0
   - G 6 *h*, 2 *t*, and 7
   - H 1 *h*, 0 *t*, and 0
   - I 7 *h*, 6 *t*, and 5

2. Give the missing digit.
   - A 384 means 3 hundreds, ▦ tens, and 4 ones.
   - B 659 means ▦ hundreds, 5 tens, and 9 ones.
   - C 518 means 5 hundreds, 1 ten, and ▦ ones.
   - D 304 means ▦ hundreds, 0 tens, and 4 ones.
   - E 927 means 9 hundreds, ▦ tens, and 7 ones.

3. Write the numeral for each part.
   - A two hundred eighty-three
   - B five hundred sixty-seven
   - C nine hundred forty-one
   - D six hundred fifty-four
   - E three hundred thirty-nine
   - F seven hundred twenty-eight

4. Find the number that is 1 more than
   - A 9
   - B 19
   - C 29
   - D 49
   - E 69
   - F 79
   - G 89
   - H 99
   - I 109
   - J 119
   - K 139
   - L 439
   - M 199
   - N 699
   - O 899

5. Copy each column and complete the counting.

   A    B

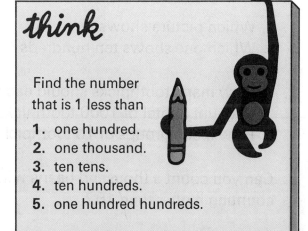

think

Find the number that is 1 less than
1. one hundred.
2. one thousand.
3. ten tens.
4. ten hundreds.
5. one hundred hundreds.

More practice, page A-5, Set 7

● *How large is a thousand?*

Investigating the Ideas

  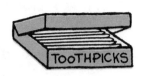

 How long do you think it would take you to count ten hundred objects? Try this with a set of objects such as one of those above.

Discussing the Ideas

 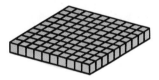 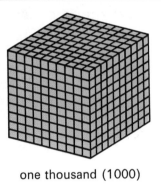

ten (10)   one hundred (100)   one thousand (1000)

1. A Which picture shows ten tens?
   B Which one shows ten hundreds?

2. A How many toothpicks should each of 10 children count to count a total of 1000 toothpicks?
   B How many groups of 10 toothpicks must each child count?

3. Can you count a thousand beans without ever counting higher than 10?

4. How many tens make a thousand?

Using the Ideas

1. Give the number for each set.
   There are 10 in each bundle and 100 in each box.

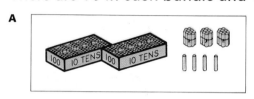

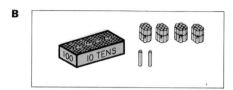

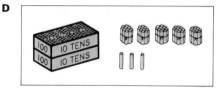

2. Give the missing numerals in the table.

We see	We think	We write
	7 hundreds	700
	8 hundreds	A
	9 hundreds	B
	10 hundreds	C

3. Write the correct numerals.
   A For 4 hundreds, 8 tens, and 3, we write ▓.
   B For 4 hundreds, 9 tens, and 7, we write ▓.
   C For 9 hundreds, 9 tens, and 7, we write ▓.
   D For 9 hundreds, 9 tens, and 8, we write ▓.
   E For 9 hundreds, 9 tens, and 9, we write ▓.
   F For 10 hundreds, 0 tens, and 0, we write ▓.

● *Let's explore 4-digit numerals.*

Investigating the Ideas

Look up one of the following.

A Price of your favorite car
B Number of feet or yards in a mile
C Number of miles from Los Angeles to New York

||||| feet = 1 mile
||||| yards = 1 mile

 Can you find some other examples of 4-digit numerals in newspapers, magazines, and books?

Discussing the Ideas

1. Study the figure below and answer the questions.

  1000   1000   1000

   A How many thousands?   C How many tens?
   B How many hundreds?   D How many ones?
       E Explain the meaning of the numeral 3462.

2. A Can you read the numeral on the sign?
   B How many thousands in the numeral?
   C How many hundreds in the numeral?
   D How many tens? ones?

CENTERVALE
6048
POPULATION

3. Read the number. Then tell how many thousands, hundreds, tens, and ones.

   A 7264   E 1635   I 9025   M 8340   Q 9216
   B 8315   F 7986   J 8840   N 2600   R 7007
   C 9126   G 8204   K 7602   O 5000   S 6000
   D 8427   H 3716   L 9100   P 8083   T 5080

Using the Ideas

1. Give the 4-digit numeral for each set.

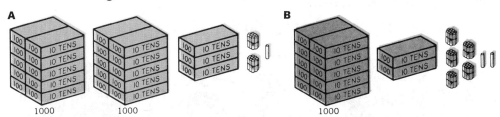

2. Write the 4-digit numeral for each of these. (*th* stands for thousands, *h* stands for hundreds, and *t* stands for tens.)
   - A 6 *th*, 5 *h*, 2 *t*, and 4
   - B 9 *th*, 4 *h*, 2 *t*, and 1
   - C 9 *th*, 4 *h*, 2 *t*, and 0
   - D 9 *th*, 4 *h*, 0 *t*, and 0
   - E 9 *th*, 0 *h*, 0 *t*, and 0
   - F 6 *th*, 0 *h*, 8 *t*, and 0

3. In each numeral below, one of the digits is red. Give the number for which that digit stands. For example, in exercise A the 7 stands for 700.
   - A 6728
   - B 4325
   - C 4286
   - D 9515
   - E 7106
   - F 8732
   - G 9457
   - H 1260
   - I 4037
   - J 5208

4. Find the missing digit for each of these.
   - A 6721 means ▨ hundreds, 6 thousands, 1 one, 2 tens.
   - B 4362 means 4 thousands, ▨ tens, 3 Hundreds, 2 ones.
   - C 7820 means 2 tens, 8 hundreds, 0 ones, ▨ thousands.
   - D 5207 means ▨ hundreds, 0 tens, 5 thousands, 7 ones.

5. Copy each column and complete the counting.

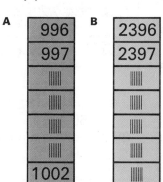

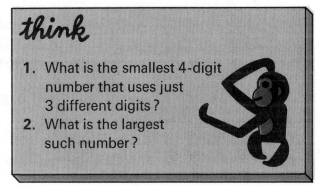

think
1. What is the smallest 4-digit number that uses just 3 different digits?
2. What is the largest such number?

More practice, page A-6, Set 8

● *Let's compare the "sizes" of numbers.*

Investigating the Ideas

Choose one of these investigations.

A Find two important dates in history.

B Find the price at two stores of something you want to buy.

C Find the numbers of your heartbeats and breaths per minute.

 Can you tell which of the two numbers you found is greater?

Discussing the Ideas

1. Study the figure. Then answer the questions.

   A Which is greater, 500 or 800?

   B Which is greater, 105 or 108?

   C Which is greater, 150 or 180?

   D Which is greater, 524 or 824?

2. Explain an easy way to remember how to use the inequality marks (< and >).

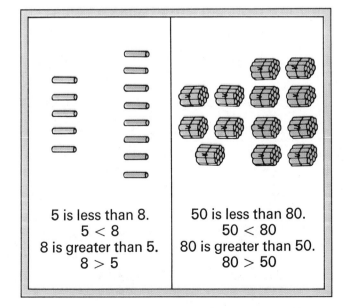

5 is less than 8.
5 < 8
8 is greater than 5.
8 > 5

50 is less than 80.
50 < 80
80 is greater than 50.
80 > 50

Using the Ideas

1. Which of the two numbers is greater?
   - A 9 or 7
   - B 5 or 8
   - C 6 or 2
   - D 80 or 20
   - E 10 or 50
   - F 60 or 40
   - G 900 or 300
   - H 400 or 700
   - I 600 or 500
   - J 2000 or 5000
   - K 6000 or 4000
   - L 8000 or 9000

2. Answer true or false for each exercise.
   - A 9 is greater than 7.
   - B 90 is greater than 70.
   - C 50 is greater than 80.
   - D 51 is greater than 50.
   - E 68 is greater than 64.
   - F 72 is greater than 73.
   - G 200 is less than 500.
   - H 300 is less than 200.
   - I 604 is less than 607.
   - J 820 is less than 850.
   - K 750 is less than 720.
   - L 930 is less than 940.

3. Which of the two numbers is greater?
   - A 8 or 3
   - B 6 or 9
   - C 5 or 8
   - D 82 or 32
   - E 67 or 97
   - F 35 or 38
   - G 68 or 63
   - H 625 or 925
   - I 655 or 685
   - J 628 or 623
   - K 6219 or 9219
   - L 4519 or 4819

4. Write each number pair on your paper. Then put the correct mark in place of the ⬤. Write the two numbers in the order given.
   - A 2 ⬤ 8   Answer: 2 < 8
   - B 80 ⬤ 50
   - C 72 ⬤ 92
   - D 72 ⬤ 71
   - E 72 ⬤ 78
   - F 200 ⬤ 500
   - G 230 ⬤ 530
   - H 837 ⬤ 537
   - I 520 ⬤ 580
   - J 684 ⬤ 654
   - K 8237 ⬤ 8537

think 9536 4871

1. Find the largest 4-digit number that uses no digit twice.
2. Find the smallest 4-digit number that uses no digit twice.
3. Find the largest 4-digit number that uses 2 digits twice each.

More practice, page A-6, Set 9

● *Do you really use large numbers?*

Investigating the Ideas

The pictures below suggest some large numbers.

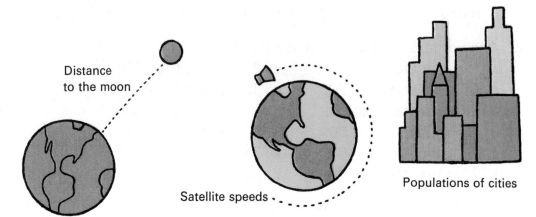

Distance to the moon

Satellite speeds

Populations of cities

 Can you find and read some numerals with 5 or more digits?

Discussing the Ideas

1. Give the number of thousands.

   A **6**287   Answer: 6
   B **27**,287
   C **394**,287
   D **9**564

   E **37**,564
   F **291**,564
   G 53,486
   H 7486

   I 875,486
   J 326,439
   K 52,475
   L 18,627

2. Write a numeral with 5 or 6 digits on the chalkboard. Have a classmate read it and tell how many thousands, hundreds, tens, and ones.

3. For 1 million, we write 1,000,000. Explain how to write
   A 2 million.
   B 3 million.
   C 23 million.
   D 672 million.

Using the Ideas

1. Give the number of thousands. For part A, write 5.
   For part B, write 38.
   - A **5**392
   - B **38**,467
   - C 7682
   - D 23,487
   - E 53,007
   - F 467,265
   - G 100,005
   - H 999,999

2. Write the numeral for each exercise.
   - A seven thousand, two hundred twenty-six
   - B fourteen thousand, five hundred eighty-three
   - C ninety-six thousand, four hundred thirty-eight
   - D one hundred twenty-six thousand, two hundred seventy-six
   - E three hundred eighty-six thousand, four hundred thirty
   - F nine hundred ninety-nine thousand, nine hundred ninety-nine

3. Give the next three numbers for each of these.

A	B	C	D
50,000	95,000	500,000	950,000
60,000	96,000	600,000	960,000
70,000	97,000	700,000	970,000
⋮	⋮	⋮	⋮

think

I think I'm big until
   I spy
So many numbers
   larger than I.
My name has a one
And zeros galore.
Seven digits in all,
   and not one more.

WHO AM I?

4. Guess the correct answer to each of these questions.
   - A How long is 1 million seconds? (about a day; about 4 days; about 2 weeks)
   - B How many trips around the world would take you 1 million miles? (about 2; almost 10; about 40)
   - C How many full-sized cars weigh 1 million pounds? (almost 10; about 75; almost 300)

Solving Story Problems
Travel Fun

Jim liked to watch the odometer on his father's car when he and his family were on vacation last summer. When they started the trip the odometer looked like this:

The odometer counts miles traveled.

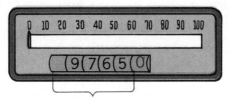

This part tells the number of miles.

1. How far had Jim's family traveled when the odometer looked like this? → (9 (7 (8 (5 (

2. Jim saw this sign. REDWOOD 100 MILES

   He looked at the odometer and saw this. → (9 (7 (9 (8 (

   What did the odometer show when they reached Redwood?

3. At Wood City the odometer looked like this. → (1 (0 (7 (8 (4 (

   At River City the odometer looked like this. → (1 (0 (7 (9 (4 (

   How far is it from Wood City to River City?

4. Logtown is 100 miles farther from Jim's house than Fish Hook is.

   The odometer looked like this at Fish Hook. → (1 (1 (0 (6 (5 (

   What do you think the odometer read at Logtown?

46

Mountain Peaks in North America

The table gives the heights of some of the highest mountains in North America. The first two mountains in the table are the highest in North America.

NAME	PLACE	FEET
McKinley	Alaska	20,320
Logan	Canada	19,850
Citlaltepetl	Mexico	18,700
King	Canada	17,130
Steele	Canada	16,440
Bona	Alaska	16,420
Wood	Canada	15,885
Bear	Alaska	14,850
Whitney	California	14,495
Elbert	Colorado	14,431
Rainier	Washington	14,410
Lincoln	Colorado	14,284

1. What mountain peak in the table is more than 17,000 feet and less than 18,000 feet?

2. What mountain peak in the table is between 15,000 feet and 16,000 feet?

3. How many peaks in the table are less than 16,000 feet?

4. How many peaks in the table are more than 15,000 feet? How many are more than 14,000 feet?

5. If an airplane flies at 17,000 feet, how many of these peaks could it fly over?

6. Which mountain peaks are more than 14,400 feet and less than 14,500 feet?

7. How much higher is Mount Steele than Mount Bona?

8. How much higher is Mount Elbert than Mount Rainier?

9. Mount Hubbard (not listed in the table) is 100 feet higher than Mount Bear. How high is Mount Hubbard?

10. How much higher is Mount Wood than Mount Bear?

Reviewing the Ideas

1. Give the number for each set of sticks.

2. Write the numeral for each of these.
   - A 7 tens and 6
   - B 3 tens and 0
   - C 5 tens and 2
   - D 6 tens and 9
   - E 8 hundreds, 6 tens, and 5
   - F 7 hundreds, 1 ten, and 2
   - G 1 hundred, 0 tens, and 9
   - H 9 hundreds, 9 tens, and 9

3. How many sticks in this set?

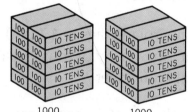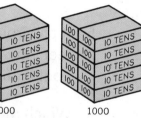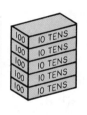

4. Write the numeral for each of these.
   - A seventy-five
   - B forty-six
   - C two hundred eighty-seven
   - D four hundred ninety-three
   - E six thousand, four hundred thirty-two
   - F twenty-three thousand, one hundred sixty-nine
   - G four hundred sixty-eight thousand, two hundred twenty-one

5. Give the next three numbers in each counting sequence.
   - A 16, 17, 18, ...
   - B 55, 56, 57, ...
   - C 96, 97, 98, ...
   - D 127, 128, 129, ...
   - E 397, 398, 399, ...
   - F 996, 997, 998, ...

6. Give the correct sign > or < for each ●.
   - A 68 ● 78
   - B 43 ● 33
   - C 127 ● 327
   - D 546 ● 526
   - E 3285 ● 3286
   - F 9463 ● 9763

Keeping in touch with   Counting
                              Measurement
                              Fractions

1. How many green strips long is the blue strip?

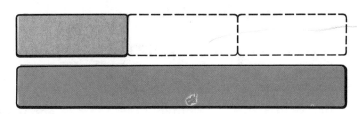

2. Use your ruler to measure the nail file to the nearest inch.

3. Measure the nail file in exercise 2 to the nearest centimeter.

4. A Is the tip of the screw nearer to 2 inches or to $1\frac{1}{2}$ inches?
   B To the nearest half inch, the length is ▓ inches.

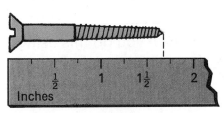

5. Give the fraction that tells what part of each region is colored.

   A   B   C   D

6. Find the area of each shaded region. Each small square is a unit.

   A 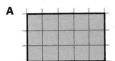   B 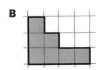   C

7. Give the volume of each figure below.

   A   B   C

 You are invited to explore   **ACTIVITY CARD 1** Page 309

3 Addition and Subtraction

• *Are addition and subtraction related?*

Investigating the Ideas

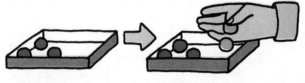

Start with 3   put in 2   and write an addition equation.

$3 + 2 = 5$

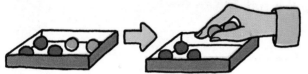

Now you have 5   take out 2   and write a subtraction equation.

$5 - 2 = 3$

 Can you do this using different numbers of counters? | Write the equations.

Discussing the Ideas

1. In each part, tell how the picture helps explain the equation.

A 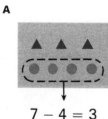   B 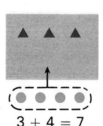   C 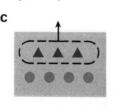   D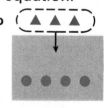

$7 - 4 = 3$ $3 + 4 = 7$ $7 - 3 = 4$ $4 + 3 = 7$

2. Write an equation for each figure.

A 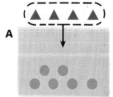   B 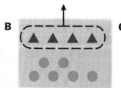   C 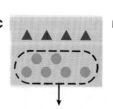   D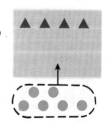

Using the Ideas

1. Write four different equations for each set.

   Example: Answer:
   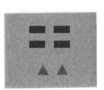 $4 + 2 = 6$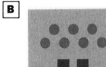
   $6 - 2 = 4$
   $2 + 4 = 6$
   $6 - 4 = 2$

2. Answer the questions about the sets.

   A How many dots in sets U and V together?
   B How many dots in sets W and Y together?
   C How many dots in sets V and Z together?
   D How many dots in sets U and Y together?
   E How many dots in sets W and X together?

3. Write an addition equation for each part of exercise 2.

4. Think about removing the dots inside the dotted ring and write a subtraction equation for each set. Then think about putting the dots back inside the dotted ring and write an addition equation for each set.

   A 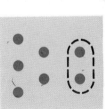 B 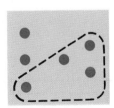 C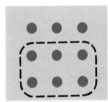

5. Write two addition and two subtraction equations using the numbers 4, 5, and 9.

6. Use 5, 3, and one other number and write two addition and two subtraction equations.

51

● Can the number line help you with addition and subtraction?

Investigating the Ideas

You can use your strips and your centimeter ruler to show addition.

$5 + 2 = 7$

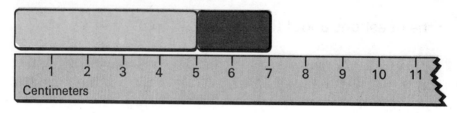

 How many other pairs of your strips can you use to add to 7?

Discussing the Ideas

1. How can you use your strips and ruler to show the subtraction equation? $7 - 2 = 5$

2. We can use **arrows** and a **number line** to show addition and subtraction instead of strips and a ruler. Give the missing number in the equation for each number line.

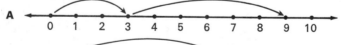

$3 + 6 = n$

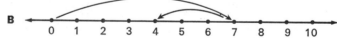

$7 - 3 = n$

3. What equation can you write for each number-line picture?

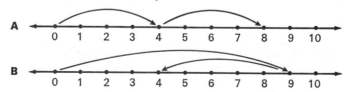

Using the Ideas

1. Write an equation for each number-line picture.

A 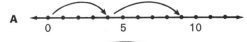   D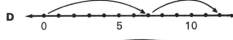

B    E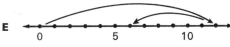

C    F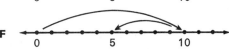

2. Draw a 12-centimeter line on your paper. Starting at the 0 dot, mark dots each 1 centimeter and label them as shown.

   A Use arrows to show 5 + 6.   B Show arrows for 11 − 9.

3. Write an equation for each part of exercise 2.

4. Write an addition equation for the picture.

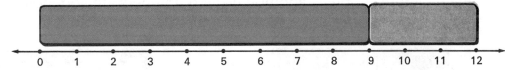

5. Draw a number line and show the sum 3 + 3 + 3 + 3.

★ 6. Write an equation for each number-line picture.

A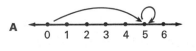

B

There is no largest number.
The last one can't be found.
Just add me to get the next.
There is no upper bound.

WHO AM I?

53

● *Let's explore addends and sums.*

Investigating the Ideas

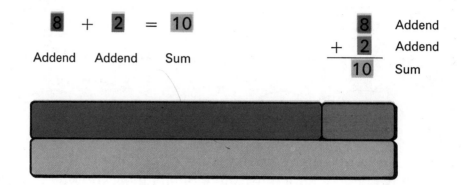

| Can you use your strips to find other pairs of addends that will give a sum of 10? | Write an equation for each pair you find. |

Discussing the Ideas

1. Each picture suggests addition. Which numbers are addends? What is the sum?

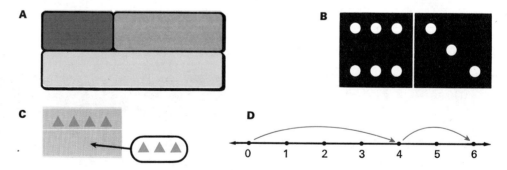

2. The sum is 8. One addend is 5. The other addend is 3. Explain how to write an addition equation for these three numbers.

3. Use 2 as an addend as many times as you need to write an equation with a sum of 10.

Using the Ideas

1. The addends are given. Find the sums.

   A) 3 + 3 B) 3 + 4 C) 3 + 2 D) 4 + 4 E) 4 + 3 F) 4 + 5 G) 5 + 5

2. Solve the equations.

   A) $5 + 4 = n$ C) $2 + 6 = n$ E) $4 + 5 = n$ G) $6 + 2 = n$
   B) $3 + 6 = n$ D) $5 + 5 = n$ F) $2 + 7 = n$ H) $8 + 2 = n$

3. Give the missing numbers in the addition tables.

Add 2	
5	7
3	5
A) 6	▨
B) 4	▨

Add 4	
4	8
C) 5	▨
D) 3	▨
E) 6	▨

Add 3	
7	10
F) 5	▨
G) 6	▨
H) 4	▨

Add 5	
I) 2	▨
J) 5	▨
K) 3	▨
L) 4	▨

4. Copy each addition table and give the missing numbers.

   A)
+	5	4
3	8	7
2	7	▨

   B)
+	4	6
4	8	▨
3	▨	▨

   C)
+	5	0
1	▨	▨
4	▨	▨

   D)
+	7	6
2	▨	▨
3	▨	▨

5. Give the missing numbers in the table.

	Addend	Addend	Sum
	3	2	5
A	5	3	▨
B	4	▨	6
C	7	▨	9
D	▨	3	5
E	3	▨	8
F	▨	4	10

think

Find two numbers so that
their sum is 9
and
their difference is 9.

● Can you find differences by finding missing addends?

Investigating the Ideas

Find three numbers from the set so that

two are **addends**
and
the other is their **sum**.

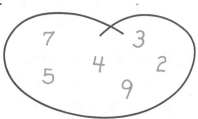

 Can you use your numbers to write two addition equations **and** two subtraction equations?

Discussing the Ideas

1. The letters A and S help you think about addends and their sum

   A A S
   $6 + 2 = 8$

   A Write another addition equation using these numbers.

   B Write two subtraction equations using these numbers.

2. The third-grade children checked their arithmetic papers. Kristy checked Jay's paper. Under one exercise, she wrote

   A Explain what Kristy was trying to tell Jay.

   B What do you think Kristy would say about these exercises?

   $8 - 3 = 4$

Using the Ideas

1. Find the missing addends.

   A $n + 4 = 7$ E $n + 4 = 8$ I $n + 0 = 8$
   B $n + 3 = 10$ F $n + 5 = 10$ J $n + 7 = 10$
   C $n + 4 = 9$ G $n + 6 = 9$ K $n + 4 = 5$
   D $n + 6 = 10$ H $n + 1 = 7$ L $n + 3 = 9$

2. Find the differences by thinking about missing addends.

   Think: $? + 4 = 7$ Think: $? + 3 = 10$ Think: $? + 4 = 9$

   A $7 - 4 = n$ B $10 - 3 = n$ C $9 - 4 = n$
   D $10 - 6 = n$ E $9 - 6 = n$ F $10 - 7 = n$
   G $8 - 4 = n$ H $7 - 1 = n$ I $5 - 4 = n$
   J $10 - 5 = n$ K $8 - 0 = n$ L $9 - 3 = n$

3. Give the missing numbers. Check your answer.

7	−3	4	+2	6	−4	A	+8	B	−1	C	−5	4
2	+6	8	+2	10	−9	D	+4	E	+2	F	−7	0

4. Find the differences. Check your answers.

   A 10   B 10   C   8
      −6   −3   −2

   D   6   E   7   F   5
      −3   −3   −5

   G   9   H   8   I   6
      −3   −1   −0

   J   9   K 10   L 10
      −5   −7   −5

think

Addition is my dearest friend.
We never fuss or fight.
When I am done,
He comes and checks
To see that I am right.

WHO AM I?

More practice, page A-7, Set 11

● Can subtraction be used to compare sets?

Investigating the Ideas

Choose one pile of 12 counters and a second pile of 7 counters. Match the counters from the piles one-to-one until one pile is used up.

 Can you copy and complete this equation about the piles? It tells how many counters were not matched. $12 - 7 = ?$

Discussing the Ideas

1. How many more caps are there than hooks? By matching we see there are |||| more caps than hooks. Here is the subtraction equation: $8 - 6 = n$

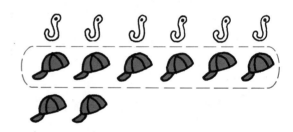

2. How many more balls are there than blocks? The matching lines show there are |||| more balls than blocks. Also, you can use subtraction. $9 - 5 = n$

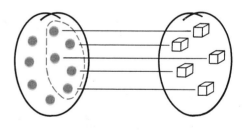

3. For each pair of sets, there are more black dots than red dots. Write a subtraction equation for each pair to tell how many more black dots.

Example:    Equation: $6 - 4 = 2$ A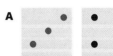

B    C   D

58

Collecting Shells

Using the Ideas

Nan collects shells. She also collects facts about shells. Page 1 in her notebook looked like this.

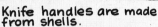

Knife handles are made from shells.
Buttons are made from shells.
Some shells are used to make roads.
Some windows are made of shells instead of glass.
Some people use shells for money.
Some shells are very pretty.

Tree-snail shells

1. Nan collected 5 tree-snail shells and 3 clam shells. How many more tree-snail shells does she have than clam shells?

2. Nan and Jill went to the beach to look for shells. Nan placed her shells in a row like this. Jill placed her shells like this.

   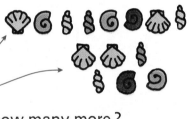

   A Who had more shells?   B How many more?

3. Nan found 9 mussel shells beside a river. She gave 4 of them to Jill. How many did Nan have left?

4. Nan wants pictures of her shells. She has 9 different snail shells. She has pictures of 7 of them. How many more pictures does she need?

5. Nan took 8 colored shells to school. She had 3 colored shells at home. She had fewer colored shells at home than at school. How many fewer?

6. Jill had 4 blue shells and 3 white shells in a box. She gave Nan 5 of these shells. How many shells are left?

More practice, page A-8, Set 12

● *What are the order and grouping principles?*

Discussing the Ideas

1. Follow a 4-strip with a 3-strip. Then follow a 3-strip with a 4-strip.

    A Do the trains match?
    B Try this again using other strips. What do you find?
    C Complete this sentence about the order principle.

    > When we change the __?__ of the __?__, we get the same sum.

2. A Do the two trains below match? Explain the grouping.

    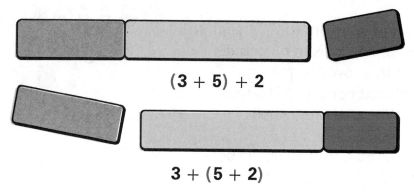

    $(3 + 5) + 2$

    $3 + (5 + 2)$

    B Addends: 4, 5, 7. The arrows tell what to add first.

    $$(4 + 5) + 7 \stackrel{?}{=} 4 + (5 + 7)$$

    Are the two sums equal?

    C Complete this sentence about the grouping principle.

    > When we change the __?__ of the __?__, we get the same sum.

Using the Ideas

1. The number line in part A shows that 2 + 5 = 5 + 2.
   What does the number line in part B show?

   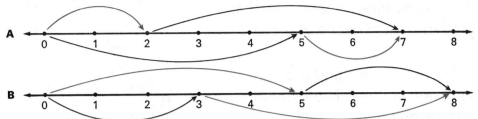

2. Solve the equations.
   A 4 + 6 = 6 + *n*
   B 5 + *n* = 3 + 5
   C *n* + 8 = 8 + 9
   D 7 + 9 = *n* + 7
   E 327 + *n* = 22 + 327
   F 9856 + 6542 = *n* + 9856

3. Give the missing number.

   IF 137 + 387 = 524   THEN 387 + 137 = *n*

4. Find the sums. Use the grouping shown by arrows.
   A {1, 5, 2}   C {1, 5, 2}   E {2, 4, 1}   G {2, 4, 1}
   B {3, 4, 2}   D {3, 4, 2}   F {5, 3, 2}   H {5, 3, 2}

5. The last two addends are grouped in the problems below. By changing the grouping, you can find the same sum more easily. Do this.
   A 7 + (3 + 5)   C 999 + (1 + 235)   E 6 + (4 + 8321)
   B 78 + (2 + 6)   D 99 + (1 + 532)   F 7 + (3 + 8267)

6. Find the sums.

   A IF (35 + 27) + 68 = 130   THEN 35 + (27 + 68) = *n*.

   B IF (56 + 39) + 28 = 123   THEN 56 + (39 + 28) = *n*.

● Let's explore rearranging addends.

Investigating the Ideas

Make three slips of paper like these.
Then turn them over and mix them up.

Pick any two slips and add the numbers on them. Then add the number on the other slip.

 If you do this five times, will you get the same sum each time?

Discussing the Ideas

1. In the Investigation you might have picked the pair ②, ④ first. What other pairs might you have picked?

2. With the three addends 2, 3, and 4, we could

A add these first.	B add these first.	C add these first.
2 3 4	2 3 4	2 3 4
(2 + 3) + 4	2 + (3 + 4)	(2 + 4) + 3

   Answer these questions for A, B, and C:
   Which two addends are grouped together?
   What is their sum? What is the total sum?

3. We can add any two numbers first. Which two would you add first in the problem 6 + 8 + 4? Why?

Using the Ideas

1. Solve these equations for addends 2 3 5.
   - A (2 + 3) + 5 = *n*
   - B (5 + 2) + 3 = *n*
   - C (3 + 5) + 2 = *n*
   - D (2 + 5) + 3 = *n*

2. However we order or group,
   - A when the addends are 2 4 3, the sum is ▥.
   - B when the addends are 4 1 2, the sum is ▥.
   - C when the addends are 2 5 0, the sum is ▥.

3. Find the sums. Look for tens.
   - A 7 + 3 + 2
   - B 7 + 2 + 3
   - C 9 + 8 + 1
   - D 6 + 2 + 4
   - E 5 + 6 + 5
   - F 4 + 6 + 9
   - G 7 + 8 + 2
   - H 5 + 7 + 3
   - I 6 + 3 + 1
   - J 5 + 2 + 8
   - K 8 + 5 + 2
   - L 9 + 5 + 5
   - M 10 + 0 + 8
   - N 2 + 5 + 2
   - O 3 + 5 + 5
   - P 1 + 6 + 9

4. Find the sums. Look for tens.

A	B	C	D	E	F	G	H
8	4	4	7	9	8	6	4
2	2	8	5	9	2	0	6
+4	+8	+2	+3	+1	+8	+4	+9

5. Find the sums. Look for tens.
   - A 2 + 8 + 3 + 4
   - B 2 + 3 + 8 + 4
   - C 9 + 5 + 1 + 5
   - D 2 + 3 + 4 + 8
   - E 4 + 6 + 3 + 3
   - F 2 + 4 + 8 + 2
   - G 4 + 3 + 3 + 6
   - H 7 + 4 + 3 + 5
   - I 7 + 6 + 2 + 4

6. Find the sums. Look for tens.

A	B	C	D	E	F	G
9	8	5	6	6	7	6
7	2	7	4	7	5	5
1	3	3	7	4	3	4
+2	+1	+2	+3	+3	+3	+5

More practice, page A-8, Set 13

• *Let's explore ways to think about sums.*

Investigating the Ideas

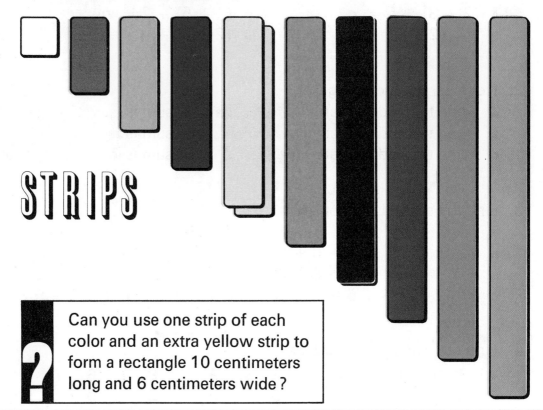

Can you use one strip of each color and an extra yellow strip to form a rectangle 10 centimeters long and 6 centimeters wide?

Discussing the Ideas

1. Give some equations suggested by rows of the rectangle.

2. A Give the missing number for the picture below.

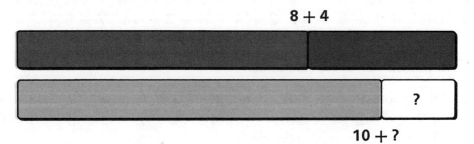

8 + 4

?

10 + ?

B Can you show some other "larger facts" using your strips?

Using the Ideas

1. Find the missing numbers. Then give the sums.
   - A $8 + 6 = 10 + n$
   - B $9 + 4 = 10 + n$
   - C $8 + 5 = 10 + n$
   - D $7 + 5 = 10 + n$
   - E $9 + 6 = 10 + n$
   - F $9 + 3 = 10 + n$
   - G $6 + 5 = 10 + n$
   - H $6 + 6 = 10 + n$
   - I $6 + 7 = 10 + n$

2. The **double** of 9 is 18 because $9 + 9 = 18$.
   Find the double of each number.
   - A 4
   - B 6
   - C 3
   - D 7
   - E 5
   - F 2
   - G 8

3. What number doubled gives each of these numbers?
   - A 10
   - B 6
   - C 14
   - D 8
   - E 18
   - F 16
   - G 2
   - H 0

4. Read each exercise carefully. Then give the sum.
   - A Because $4 + 4 = 8$, we know that $4 + 5 = n$.
   - B Because $7 + 7 = 14$, we know that $7 + 6 = n$.
   - C Because $5 + 5 = 10$, we know that $5 + 6 = n$.
   - D Because $6 + 6 = 12$, we know that $6 + 7 = n$.
   - E Because $8 + 8 = 16$, we know that $8 + 9 = n$.

5. Find the sums.
   - A 2 +8
   - B 6 +3
   - C 9 +4
   - D 7 +6
   - E 8 +1
   - F 4 +6
   - G 7 +2
   - H 9 +8
   - I 7 +7
   - J 2 +7
   - K 8 +3
   - L 7 +8

I'm quite a tiny number.
So very small indeed.
If you add me to another,
A change you cannot read.

WHO AM I?

6. Give the sums.
   - A 3, 4, +5
   - B 5, 3, +7
   - C 5, 1, +9
   - D 5, 4, +6

More practice, page A-9, Set 14

● *How can addition help you find differences?*

Discussing the Ideas

You can find differences if you can find missing addends.

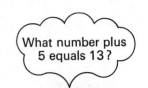

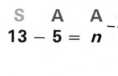
S   A   A
13 − 5 = *n*

1. When the sum is 13 and one addend is 5, how can you find the other addend?

2. Read each sentence aloud and give the difference.

       A A S   S A A
  A Because 6 + 7 = 13, we know that 13 − 7 = *n*.
  B Because 8 + 7 = 15, we know that 15 − 7 = *n*.
  C Because 9 + 5 = 14, we know that 14 − 5 = *n*.
  D Because 8 + 6 = 14, we know that 14 − 6 = *n*.

3. Give the difference and explain how you know it is correct.
  A Because 48 + 37 = 85, we know that 85 − 37 = *n*.
  B Because 76 + 88 = 164, we know that 164 − 88 = *n*.
  C Because 57 + 19 = 76, we know that 76 − 19 = *n*.

4. Give the missing number and explain how it helps you find the difference.
  A To find 15 − 7, it helps to think *n* + 7 = 15.
  B To find 17 − 9, it helps to think *n* + 9 = 17.
  C To find 14 − 6, it helps to think *n* + 6 = 14.
  D To find 12 − 7, it helps to think *n* + 7 = 12.

Using the Ideas

1. Find the differences.
   - A $12 - 4 = n$
   - B $11 - 5 = n$
   - C $13 - 6 = n$
   - D $14 - 5 = n$
   - E $14 - 7 = n$
   - F $15 - 8 = n$
   - G $16 - 9 = n$
   - H $12 - 6 = n$
   - I $11 - 8 = n$

2. Find the sums and differences. Use any method you choose.

A 8 +2	B 9 −6	C 8 +7	D 7 +6	E 12 −3	F 14 −6	G 7 +8	H 8 +0
I 13 −4	J 13 −5	K 6 +6	L 17 −7	M 18 −9	N 6 +8	O 5 +5	P 7 +3
Q 10 −4	R 10 +8	S 9 +7	T 15 −7	U 8 +8	V 9 +9	W 16 −6	X 14 −9

Short Stories

1. 5 merit badges. Need 14 in all. How many more needed?

2. 13 balloons. Stick 8 with a pin. How many left?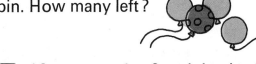

3. 16 flies. 7 frogs. Each frog gets a fly. How many flies are left?

4. 12 peanuts. Ate 2 and drank milk. Ate 5 more. How many left?

5. Caught 17 fish. 9 too small so threw them back. How many left?

6. Caught 12 butterflies. 7 got away. How many left?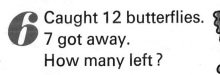

7. 17 papers. Want to sell all but 10. How many must be sold?

8. Gave 8 valentines. Received 17 valentines. Gave how many fewer than received?

More practice, page A-10, Set 15

● *Let's play "What's My Rule."*

Investigating the Ideas

Tom thought of a rule.
When Jane said 3, Tom answered 5.
When Jane said 8, Tom answered 10.
When Jane said 5, Tom answered 7.
When Jane said 2, Tom answered 4.
Jane tried to figure out what rule
Tom was using.

Dick thought of a rule.
When Sue said 2, Dick answered 4.
When Sue said 5, Dick answered 10.
When Sue said 8, Dick answered 16.
Sue said 9. Dick answered 18.
Sue thought, "What did Dick do to my number to get his number?"

What rules were the boys using?

 Make up a rule of your own and play "What's My Rule" with a classmate. Can you guess each other's rules?

Discussing the Ideas

1. How would you keep a record of what happened in a game of What's My Rule?

2. Show a record of an imaginary game, but leave out the last answer. Can your classmates give the answer?

Using the Ideas

Study the tables carefully. Guess the rule, then give what you think should go in each ▦.

1.

	Carol's number	Jill's answer
	4	14
	6	16
	2	12
A	7	▦
B	9	▦
C	▦	13
D	24	▦

2.

	Cindy's number	Nan's answer
	8	1
	7	0
	10	3
A	9	▦
B	▦	6
C	▦	8
D	▦	10

★ 3.

	Cathy's number	Susan's answer
	2	0
	4	0
	3	1
	5	1
	8	0
A	6	▦
B	11	▦

4. Choose a rule and make a table to show what might happen when you play What's My Rule.

5. Jane and Brenda played a game. Try to solve their puzzles.

  A I'm thinking of a number. If you add 6 to it, you get 10. What is the number?

  B I'm thinking of a number. If you add it to 4, you get 8. What is the number?

  C I'm thinking of a number. If you add 3 to it and then add 2, you get 10. What is the number?

  D I'm thinking of a number. If you subtract 5 from it, you get 10. What is the number?

  E If you subtract 6 from a number and then subtract 2, you get 8. What is the number?

  ★ F I'm thinking of a number. If you add it to itself and then add 4, you get 10. What is the number?

My number plus 6 equals 10

● *How does the function machine work?*

Discussing the Ideas

1. Study the pictures and explain how you think the function machine works. A record of the operations of the function machine is shown below. What are the missing numbers?

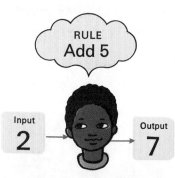

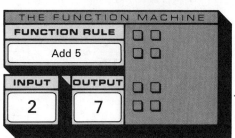

2. Find each output number. Explain how the function machine is like the student playing the What's My Rule game.

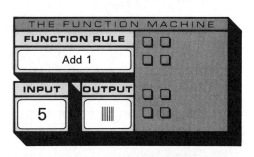

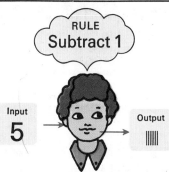

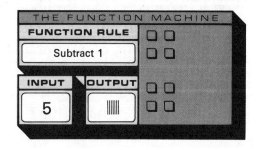

Using the Ideas

Think about the function machine and tell what you think should go in each gray space.

1. Function Rule: Add 5

	Input	Output
	2	7
	4	9
A	1	
B	5	
C	3	

2. Function Rule: Subtract 3

	Input	Output
A	6	
B	8	
C	10	
D	3	
E	13	

3. Function Rule: Add 4

	Input	Output
A	2	
B	4	
C	7	
D	8	
E		13

4. Function Rule: Subtract 2

	Input	Output
	3	1
A	7	
B		3
C	10	
D		10

5. Function Rule: Subtract 8

	Input	Output
	10	2
A	13	
B	15	
C		3
D	17	

6. Function Rule: If **odd**, subt. 1 If **even**, add 0

	Input	Output
	3	2
	5	4
A	9	
B	8	

7. Function Rule:

	Input	Output
	7	1
	10	4
	6	0
B	9	
C		8

8. Function Rule:

	Input	Output
	2	10
	5	13
	7	15
B	3	
C		17

9. Function Rule:

	Input	Output
	0	9
	1	10
	9	18
B	7	
C	5	

Reviewing the Ideas

1. Write 2 addition and 2 subtraction equations for each set.

   A    B    C

2. Find the sums and differences.

   A 8   B 6   C 9   D 2   E 10   F 7   G 6   H 4
     +2   +3   −4   +5   −4   −3   +4   +4

   I 7   J 8   K 0   L 8   M 10   N 4   O 10   P 3
     −5   −5   +6   −2   −6   +5   −3   +5

3. Find the differences.
   A $7 - 3 = n$ E $12 - 9 = n$
   B $13 - 9 = n$ F $16 - 8 = n$
   C $11 - 3 = n$ G $13 - 6 = n$
   D $15 - 7 = n$ H $14 - 7 = n$

4. Find the sums. Look for a sum of 10.
   A $6 + 4 + 7$ E $4 + 6 + 5$
   B $8 + 7 + 3$ F $7 + 8 + 3$
   C $8 + 9 + 2$ G $6 + 2 + 8$
   D $5 + 7 + 5$ H $9 + 8 + 1$

5. Find the missing numbers.
   A If $47 + 38 = 85$, then $85 - 38 = n$.
   B If $92 - 18 = 74$, then $74 + 18 = n$.
   C $27 + 68 = 68 + n$.
   D $(9 + 27) + 68 = 39 + (n + 68)$

6. Use the number line to help you solve each equation.

   A
   $7 + n + 2 = 12$

   B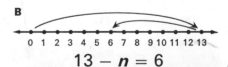
   $13 - n = 6$

   think

   Increase me by 5.
   Then take away 7.
   When you are done,
   You should have 11.

   WHO AM I?

Keeping in Touch with — Addition, Place value, Inequalities, Fractions

1. Write the missing numerals.
   A For 3 tens and 6, we write ||||.
   B For 5 tens and 0, we write ||||.
   C For 7 tens and 4, we write ||||.
   D 46 means |||| tens and 6.

2. Give the missing words.
   A In 1847, the 4 means four __?__.
   B In 6253, the 6 means six __?__.
   C In 2584, the 5 means five __?__.
   D In 3475, the 7 means seven __?__.

3. Solve the equations.
   A $63 = 60 + n$
   B $10 + 7 = n$
   C $18 = n + 8$
   D $70 = 70 + n$
   E $80 + 0 = n$
   F $30 + 4 = n$

4. Write a fraction for the part of each region that is shaded.
   A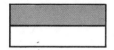
   B
   C

5. In each pair, which number is larger?
   A 624 / 621
   B 283 / 300
   C 4284 / 4384
   D 9999 / 10,000
   E 26,354 / 26,364

★ 6.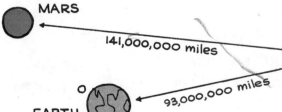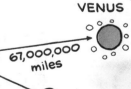

   A Which of these planets is farthest from the sun?
   B Which of these planets is closest to the sun?

 You are invited to explore — ACTIVITY CARD 2, Page 310

4 Geometry

● *Let's count edges, faces, and vertices.*

Investigating the Ideas

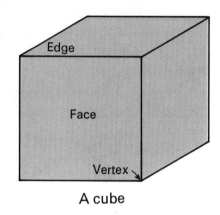
A cube

A cube has "straight" **edges**, "flat" **faces**, and "pointed" **vertices**.

 Can you find how many of each?

Discussing the Ideas

1. Can you name another object that has edges, faces, and vertices?

2. A **sphere** (ball) has no edges and no vertices. It has a **"curved" surface**. Can you name another object that has no edges or vertices?

A sphere

3. A **cone** has a **curved edge**.
   A How many vertices and flat faces does it have?
   B Can you name another object with a curved edge?

A cone

Using the Ideas

1. Name at least one figure above which has

   A no edges.
   B only straight edges.
   C only curved edges.
   D both straight and curved edges.
   E no vertices.
   F only one vertex.
   G more than one vertex.
   H only flat faces.
   I a curved face or surface.
   J both flat and curved surfaces.
   K curved edges and both flat and curved surfaces.

★ 2. Which figure has **more than one vertex**, both **flat** and **curved surfaces**, and both **straight** and **curved** edges?

● *Can you name the simplest geometric figures?*

Discussing the Ideas

1. These figures suggest points • .
   Can you think of others?

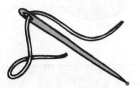  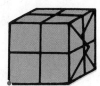

2. These figures suggest line segments •——• .
   Can you think of others?

3. A beam of light suggests a
   ray •——→ . A ray has one endpoint
   and "goes on and on" in one direction.
   Can you think of other examples of rays?

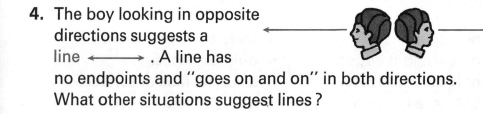

4. The boy looking in opposite
   directions suggests a
   line ←——→ . A line has
   no endpoints and "goes on and on" in both directions.
   What other situations suggest lines?

5. These pictures suggest

   planes ▱ .
   Can you think of others?

Using the Ideas

1. Mark a **point** on your paper. Use your ruler to draw 5 different **lines** that pass through that point.

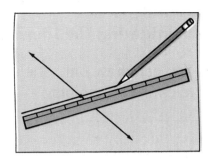

2. Mark a point on your paper. Use your ruler to draw 3 different **rays** from that point.

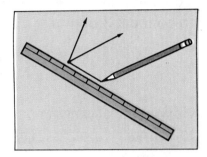

3. A Mark two points on your paper. Use your ruler to draw **the line** that passes through these two points.

   B Mark two other points on your paper. Draw **the line segment** for these points.

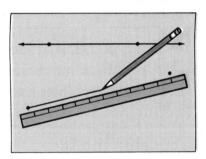

★ 4. Study the chart. Then draw and name a line, a ray, and a segment.

We see the figure	We label some points	We write a name for the figure	We say
←———→	A B	$\overleftrightarrow{AB}$	"line AB"
•———→	P Q	$\overrightarrow{PQ}$	"ray PQ"
•———•	C D	$\overline{CD}$	"segment CD"

77

● Let's count line segments.

Investigating the Ideas

You can draw only one segment to connect 2 points.

You can draw three segments to connect 3 points.

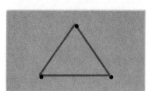

How many segments can you draw to connect 4 points? Try it.

 If 5 points are placed like this, how many segments can you draw to connect them?

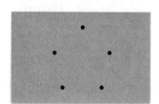

Discussing the Ideas

1. When 4 points are placed like this, how many segments do you think you can draw to connect them?

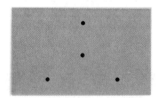

2. This chart shows the number of segments that connect each set of points shown.

Points	• •	• • •	• • • •	• • • • •	• • • • • •
Segments	1	3	6	|||||	|||||

   A How many segments did you get for 5 points?

   B Guess how many segments you get for 6 points. Try it.

Using the Ideas

1. The red segment from A to B ($\overline{AB}$) is shorter than the blue segment from C to D ($\overline{CD}$). Use the picture of the nailboard for these questions.

   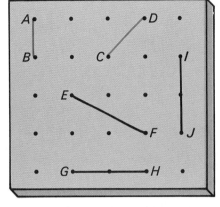

   A Which segment is the longest?
   B Which segment is shortest?
   C Can you name a segment that is longer than $\overline{AB}$ but shorter than $\overline{GH}$?
   D Name two segments that have the same length.

2. How many segments of different lengths can you show using only nine points like these?

   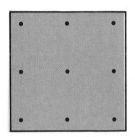

3. Trace dots A, B, C, D, and E and connect them in this order.

   A B C D E A

   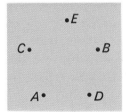

   A What figure did you make?
   B How many segments did you draw?

4. A Draw a figure using 4 segments.
   B Draw a figure using 5 segments.

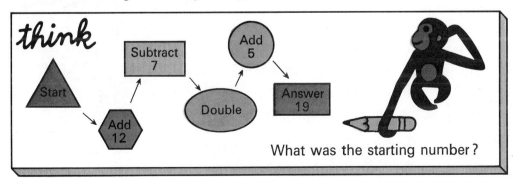

What was the starting number?

● What is an angle?

Investigating the Ideas

All of these objects are alike in some way.

 Which of these objects are like the ones above?

Discussing the Ideas

1. The compass suggests the idea of an **angle**. An angle is two rays with the same endpoint. Can you think of other things that suggest angles?

2. Draw an angle using the corner of your tablet. This special angle is called a **right angle**. Can you find some objects that suggest the idea of a right angle?

3. Can you find a way to fold a piece of paper so that some right angles are formed?

Using the Ideas

1. Follow these steps to form a right angle.

   A Fold your paper any way you wish.

   B Push your pencil point through both pieces of the folded paper.

   C Open the paper and draw a line segment connecting the two holes in your paper.

   D Use a corner of your tablet to see if the fold line and the line you drew form a right angle.

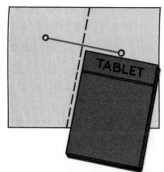

2. A Mark one point on your paper. Draw two rays from this point. The figure you have drawn shows an angle.

   B Draw four more angles. Make each one look different.

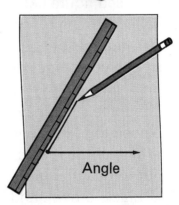
Angle

★ 3. Study the chart. Then name the angles you drew in exercise 2.

We see the angle	We label some points	We write a name for the figure	We say
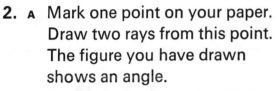	A B C	∠ABC or ∠CBA	"angle ABC" or "angle CBA"

• *What is a triangle?*

Investigating the Ideas

Two three-sided figures of different shapes are shown on the 3-by-3 geoboard.

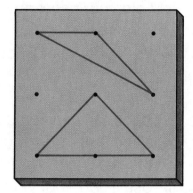

 How many three-sided figures of different shapes can you draw on 3-by-3 dot paper?

Discussing the Ideas

1. A closed figure like this one is called a triangle.
    A. A triangle has __?__ line segments.
    B. Can you name some objects that are in the shape of triangles?

2. The three strips have their corners placed together to form a triangle. Which of these sets of strips will not form a triangle?
    A. black, brown, and blue
    B. yellow, orange, and light green
    C. three yellow strips
    D. red, yellow, and brown

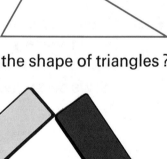

3. Suppose you drew a triangle using your light green, purple, and yellow strips and a classmate used the same strips to form a triangle. Do you think the two triangles would have the same size and shape?

Using the Ideas

1. Mark 3 points as in the figure. Draw line segments to connect those points.
   - A How many segments did you draw?
   - B What is the name of the figure you have drawn?

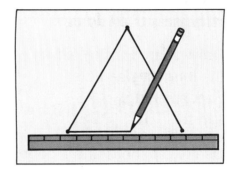

2. Draw four triangles. Make each one different.

3. Draw a triangle and mark a point inside the triangle. Put your pencil on the point and draw a path that crosses a side of the triangle.
   - A Where is your pencil point now, **inside** or **outside**?
   - B Cross again. Where is the pencil point now?
   - C If you cross 5 times in all, where are you?
   - D If you cross 8 times in all, where are you?

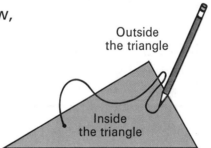

★ 4. Study the chart. Then name the triangles you drew in exercise 2.

We see the triangle	We label some points	We write a name for the triangle	We say
(triangle)	(triangle with A, B, C labeled)	△ ABC	"triangle ABC"

● Let's explore the angles of a triangle.

Investigating the Ideas

1. Draw a large triangle. Make the longest side at least 7 inches long.

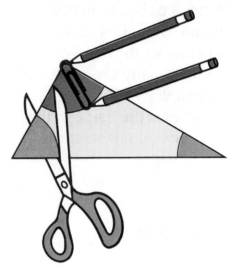

2. Use two pencils and a paper clip to draw part of a circle at each corner. Color each corner a different color. Now cut out the triangle and cut off the colored corners.

 Draw a circle with your paper clip and pencils. How much of the circle do the three corners of the triangle fill if the edges touch but do not overlap?

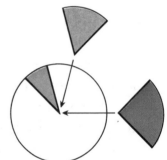

Discussing the Ideas

1. How much of a circle can you fill with corners from two triangles?

2. A triangle that has one right angle is called a **right triangle**. Do you think the corners of a right triangle will fill half a circle?

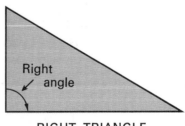
RIGHT TRIANGLE

3. Can a triangle have two right angles? Explain.

Using the Ideas

1. Use the corner of your tablet or crayon box and draw a **right triangle** on your paper.

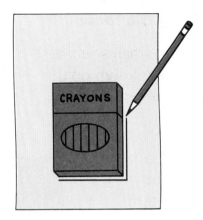

2. Draw a right triangle that has two sides that are the same length.

3. Use the corner of a sheet of paper to help you decide which of these are right triangles.

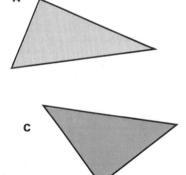
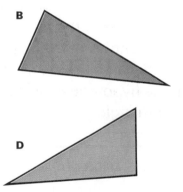

★ 4. Draw a right triangle and color the two angles that are not right angles different colors. Cut out the triangle and then cut off the colored corners. Will they fit exactly into the right angle without overlapping?

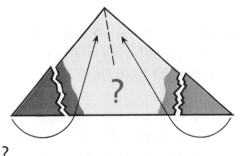

● *Let's explore some right triangles.*

Investigating the Ideas

Trace the two small squares that have been drawn on the two legs of the right triangle.

Color and cut out the four small right triangles that form the two squares.

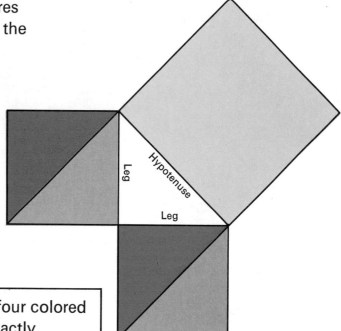

? Can you make the four colored right triangles fit exactly into the large yellow square on the hypotenuse of the right triangle?

Discussing the Ideas

1. In what way are right triangles different from other triangles?

2. Which one of your strips fits in the dotted outline to complete the right triangle?

3. What strip could you use with your 6-strip and 8-strip to form a right triangle?

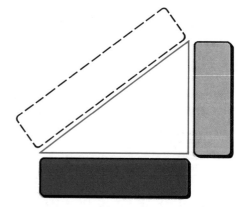

86

Using the Ideas

Use the four colored right triangles you cut out for the Investigation. Arrange the four triangles so that they exactly fit each figure. Draw a picture to show how you arranged the triangles for each figure.

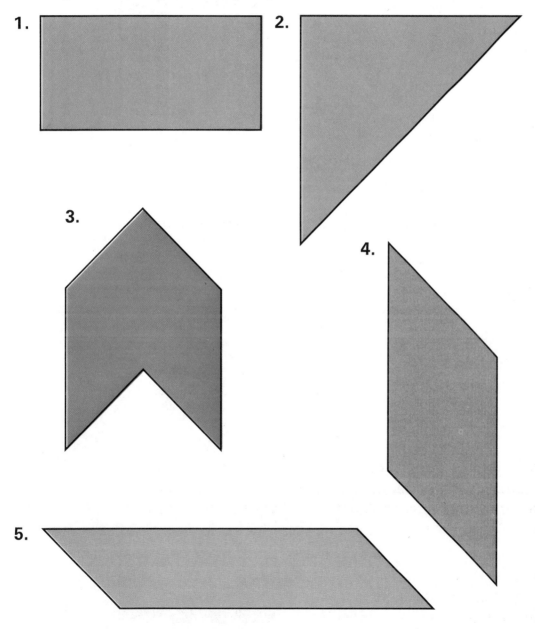

87

Reviewing the Ideas

1.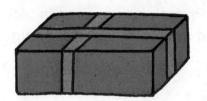
   A How many vertices (corners) does the box have?
   B How many edges does the box have?
   C How many faces does the box have?

2. A Mark three points like A, B, and C on your paper.
   B Draw a ray starting at A and going through B. Draw a ray starting at A and going through C.
   C The name for the figure is __?__. (ray, angle, triangle)

   A•
   •B
   C•

3. Which angle is a **right angle**?

   A   B   C

4. Which triangle is a **right triangle**?

   A   B   C

5. Trace these six points on your paper. How many segments can you draw to connect pairs of points?

6. How many triangles are in the picture?

   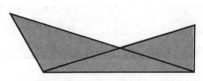

Keeping in Touch with   Measurement   Subtraction
                                      Place value   Fractions
                                      Addition

1.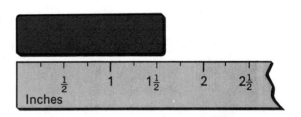
   A Is the length of the strip closer to 1 inch or to 2 inches?
   B Is the strip nearer to $1\frac{1}{2}$ inches or to 2 inches?
   C To the nearest half inch, the strip is ▨ inches long.

2. Using the unit shown, give the area of each region.

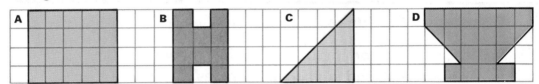

3. Write the numeral for each.
   A 4 tens and 7 ones
   B 8 hundreds, 9 tens, and 6 ones
   C seven hundred sixty-three
   D six thousand, two hundred eight

4. Write an equation for each number-line picture.

   A 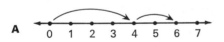   B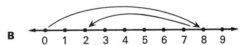

5. Find the missing addends.

   A $n + 2 = 7$ C $3 + n = 10$ E $n + 1 = 6$
   B $n + 5 = 9$ D $6 + n = 9$ F $3 + n = 8$

6. Find the differences.

	A	B	C	D	E	F
	9	8	10	6	9	10
	−7	−6	−2	−6	−8	−3

   *You are invited to explore*   **ACTIVITY CARD 3** Page 310

5 Adding and Subtracting

● *How are dimes and pennies like tens and ones?*

Investigating the Ideas

 How many different amounts of money can you show if your "pennies" and "dimes" are these different-colored counters?

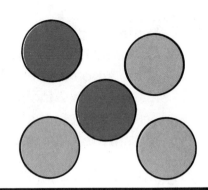

Discussing the Ideas

1. Why is it important to decide which counters are "dimes" and which are "pennies"?

2. If these counters are worth 37¢, which colored counters are dimes?

3. Which collection is worth more, A or B?

4. Which collection is worth more, C or D?

Using the Ideas

1. Give the value of each of the two coin collections together.
   - A A and B
   - B A and C
   - C C and D
   - D B and D
   - E B and C
   - F A and D

2. Which pair of collections has the greater value?
   - A A and B together or C and D together
   - B A and C together or B and D together
   - C A and D together or B and C together
   - D A and D together or B and D together

3.
   - A Had 1 dime and 7 pennies.
        Spent 5 cents.
        How much left?
   - B Had 2 dimes and 3 pennies.
        Spent 1 dime and 3 cents.
        How much left?
   - C Had 6 dimes and 7 pennies.
        Spent 5 dimes and 5 cents.
        How much left?
   - D Had 8 dimes and 2 pennies.
        Spent 40 cents.
        How much left?
   - E Had 5 dimes and 9 pennies.
        Spent 50 cents.
        How much left?
   - F Had 4 dimes and 8 pennies.
        Spent 25 cents.
        How much left?
   - G Had 76 cents.
        Spent 56 cents.
        How much left?
   - H Had 79 cents.
        Spent 43 cents.
        How much left?

★ 4. How much more is in one collection than the other?

● *How is 20 + 30 like 2 + 3?*

Investigating the Ideas

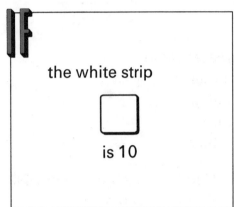

the white strip

is 10

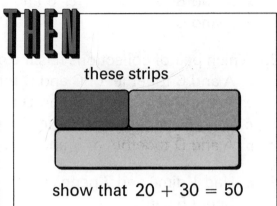

these strips

show that 20 + 30 = 50

? Can you use your strips to help you write some equations that have a sum of 80?

Discussing the Ideas

1. A 4 dimes together with 3 dimes make how many dimes?

   B 4 tens and 3 tens make how many tens?
   C 4 + 3 = *n*   D 40 + 30 = *n*

2. A 5 bundles together with 4 bundles make how many bundles?

   B 5 tens and 4 tens make how many tens?
   C 5 + 4 = *n*   D 50 + 40 = *n*

Using the Ideas

1. Find the sums.
   A. Since $7 + 2 = 9$, we know that $70 + 20 = n$.
   B. Since $2 + 5 = 7$, we know that $20 + 50 = n$.
   C. Since $6 + 4 = 10$, we know that $60 + 40 = n$.
   D. Since $7 + 5 = 12$, we know that $70 + 50 = n$.
   E. Since $8 + 7 = 15$, we know that $80 + 70 = n$.

2. Solve the equations.
   A. $60 + 10 = n$
   B. $50 + 40 = n$
   C. $40 + 30 = n$
   D. $30 + 30 = n$
   E. $50 + 60 = n$
   F. $70 + 60 = n$

3. Solve the equations.
   A. $30 + 0 = n$
   B. $30 + 1 = n$
   C. $30 + 2 = n$
   D. $30 + 7 = n$
   E. $40 + 8 = n$
   F. $60 + 4 = n$

4. Find the sums.
   A. $30 + 60$
   B. $40 + 20$
   C. $70 + 20$
   D. $70 + 40$
   E. $80 + 50$

5. Find the sums.
   A. $2 + 3 + 4$
   B. $2 + 7 + 1$

6. Find the sums.
   A. $70 + 20 + 50$
   B. $60 + 40 + 80$
   C. $20 + 30 + 20 + 10$
   D. $40 + 20 + 10 + 50$

think

With 2 minutes to play, this was the score.

| Lincoln School | 48 |
| Jefferson School | 41 |

At the end of the game, the score was 50 to 47.

WHO WON THE GAME?

● Let's explore sums and differences.

Discussing the Ideas

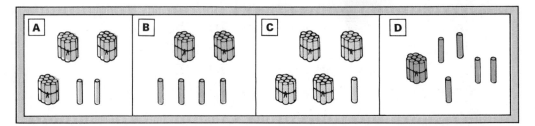

1. How many sticks are in these sets? Explain your answers.
   - A A and B together
   - B B and C together
   - C A and C together
   - D C and D together
   - E A and D together
   - F B and D together

2. Find the sums.

   A 80   7 → 87
      +10 +2   +12

   B 30   4 → 34
      +50 +2   +52

3. Give an easy rule for finding this sum.

   24
   +35

4. A How many sticks in all?
   B How many sticks in the dotted ring?
   C How many sticks not in the dotted ring?

   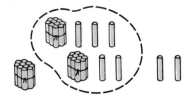

5. Find the differences.

   A 70   6 → 76
      −20 −3   −23

   B 80   5 → 85
      −30 −2   −32

6. Give an easy rule for finding this difference.

   67
   −42

Using the Ideas

1. Find the sums.

   A. 15 + 42
   B. 47 + 31
   C. 32 + 65
   D. 41 + 23
   E. 82 + 14
   F. 60 + 70
   G. 12 + 73

   H. 75 + 53
   I. 84 + 12
   J. 24 + 65
   K. 95 + 43
   L. 235 + 162
   M. 740 + 36
   N. 153 + 842

2. Find the differences.

   A. 78 − 32
   B. 93 − 41

   C. 100 − 90
   D. 82 − 62

   E. 648 − 325
   F. 739 − 516

   G. 527 − 524
   H. 607 − 403

Tom had 10 marbles. Bill said, "If you give me 2 of your marbles, then we'll both have the same number."

How many marbles did Bill have?

3. Jane checked out a library book on the thirteenth day of the month. She had to return the book in 14 days. What day was Jane's book due?

4. Tim started reading a science book on page 62. After he read 25 pages, where was he in the book?

5. Sue kept a record of the number of days it took her to read a book. How many days in all did it take for Sue to read books **A**, **B**, and **C**?

   Book A ――――12
   Book B ――――14
   Book C ――――13

More practice, page A-11, Set 16

Keeping in Touch with

Addition
Place value
Inequalities

1. Solve the equations.
   - A $38 = 30 + n$
   - B $56 = 50 + n$
   - C $70 + 2 = n$
   - D $80 + 3 = n$
   - E $358 = 300 + 50 + n$
   - F $463 = 400 + 60 + n$
   - G $781 = 700 + 80 + n$
   - H $640 = 600 + 40 + n$

2. Give the number of each set.

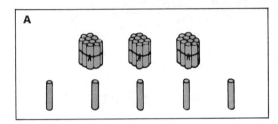
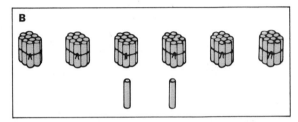
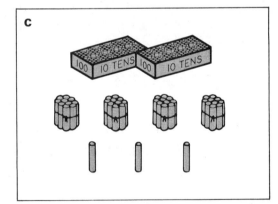
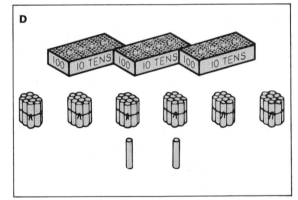

3. Solve the equations.
   - A $76 = 70 + n$
   - B $76 = 60 + n$
   - C $42 = 40 + n$
   - D $42 = 30 + n$
   - E $91 = 90 + n$
   - F $91 = 80 + n$
   - G $50 = 50 + n$
   - H $50 = 40 + n$
   - I $62 = 60 + n$
   - J $62 = 50 + n$
   - K $24 = 20 + n$
   - L $24 = 10 + n$

4. Write each number pair on your paper in the order given. Then put the correct mark (< or >) in place of the ●.

   A 65 ● 13
   (Answer: 65 > 13)
   B 27 ● 95
   C 38 ● 48
   D 55 ● 56
   E 615 ● 605
   F 82 ● 92
   G 45 ● 54
   H 69 ● 70
   I 743 ● 733

think

The sum of two numbers is 20. Their difference is 4. What are the numbers?

SUM IS 20
DIFFERENCE IS 4

5. Give the correct mark for each ●.
   A Since 7 + 5 > 10, we know that 67 + 5 ● 70.
   B Since 5 + 4 < 10, we know that 15 + 4 ● 20.
   C Since 8 + 6 > 10, we know that 48 + 6 ● 50.
   D Since 8 + 4 > 10, we know that 48 + 4 ● 50.

6. Tell whether each sum is less than 50, more than 50, or equal to 50.
   A 46 + 3   C 46 + 5   E 47 + 1   G 47 + 3
   B 46 + 4   D 46 + 6   F 47 + 2   H 47 + 4

7. Tell whether each sum is less than, more than, or equal to 70.
   A 68 + 4   C 69 + 1   E 67 + 5   G 64 + 6
   B 62 + 5   D 68 + 1   F 63 + 6   H 64 + 7

You are invited to explore

ACTIVITY CARD 4
Page 311

● How can you find sums like 36 + 28?

Investigating the Ideas

Can you use the calendar to help you find sums?
- A What is the date 3 days after your birthday?
- B Find the date 8 days after your birthday.

S	M	T	W	T	F	S
		1	2	3	4	5	6
7	8	9	10	11	12	13
14	15	16	17	18	19	20
21	22	23	24	25	26	27
28	29	30	31			

Can you use the calendar to find these sums?

23	16	24	17	19	23
+7	+6	+7	+9	+11	+8

Discussing the Ideas

1. Study the example below. Explain each step.

Step 1	Step 2	Step 3
37 +25 ――― 12	37 +25 ――― 12 50	37 +25 ――― 12 50 ――― 62
7 + 5 = 12	30 + 20 = 50	12 + 50 = 62

2. Now try this one and compare your answer with the answer your teacher puts on the chalkboard.

65
+17

Using the Ideas

1. Find the sums.

A 46 +28	B 37 +44	C 29 +65	D 67 +15	E 48 +6
F 34 +27	G 68 +19	H 17 +35	I 54 +9	J 37 +27
K 46 +35	L 27 +67	M 58 +14	N 42 +49	O 36 +27

2. Find the sums.

A 76 +40	B 76 +45	C 67 +60	D 67 +66	E 56 +56
F 83 +30	G 83 +39	H 56 +75	I 88 +33	J 95 +28
K 76 +72	L 66 +84	M 97 +14	N 59 +79	O 88 +88

3. Study the example. Then try to find the other sums without pencil and paper.

Example: 58 + 24

First
Think
58
+20
78

Then
Think
78
+4
82

A 45 + 36   C 63 + 29
B 37 + 25   D 59 + 27

think

I can be found
Halfway between
Twenty-seven
And seventeen.

WHO AM I?

17 ←——————→ 27

More practice, page A-11, Set 17

99

● *Is there a shortcut for adding with regrouping?*

Discussing the Ideas

1. Explain the difference in the way John and Susan started their work.

2. Explain the two steps in this example.

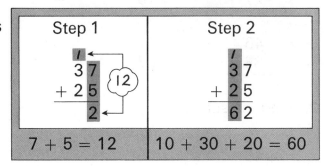

3. One of these examples has a mistake. See if you can find it.

　A 45　　B 34　　C 29
　 +38　　 +48　　 +16
　 ─── 　　 ─── 　　 ───
　　83　　　72　　　45

4. Explain each step in this example.

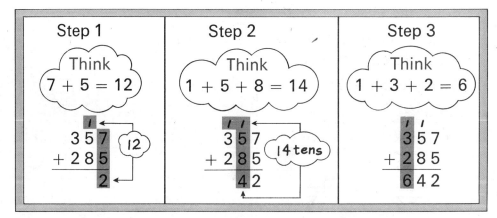

5. Try these two exercises and check your work with your teacher.

　A 248　　B 657
　 +175　　 +195

Using the Ideas

1. Find the sums.

   A. 57 + 15   B. 18 + 72   C. 49 + 33   D. 28 + 77   E. 39 + 44   F. 36 + 26

   G. 76 + 44   H. 85 + 77   I. 98 + 33   J. 84 + 27   K. 75 + 58   L. 83 + 75

   M. 85 + 26   N. 73 + 98   O. 46 + 38   P. 75 + 83   Q. 99 + 24   R. 86 + 67

2. Find the sums.

   A. 258 + 137   B. 258 + 167   C. 258 + 187   D. 546 + 125   E. 546 + 175   F. 546 + 675

   G. 546 + 237   H. 546 + 287   I. 123 + 248   J. 193 + 248   K. 648 + 176   L. 435 + 389

3. Find the sums.

   A. 804 + 307   B. 906 + 218   C. 937 + 105   D. 893 + 248   E. 346 + 928   F. 769 + 452

---

think

My hundreds' place is 9 less than
My tens and ones combined.
Twice my ones will give my tens.
My name you now can find.

WHO AM I?

HUNDREDS   TENS   ONES

● Can you find the cost of a lunch?

Investigating the Ideas

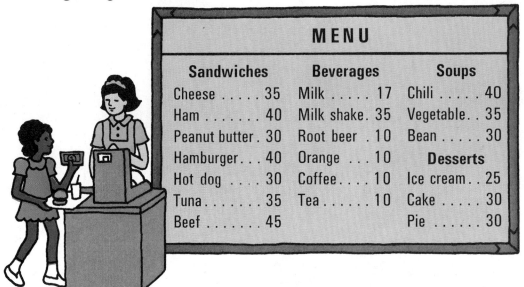

MENU		
Sandwiches	**Beverages**	**Soups**
Cheese 35	Milk 17	Chili 40
Ham 40	Milk shake . 35	Vegetable . . 35
Peanut butter . 30	Root beer . 10	Bean 30
Hamburger . . . 40	Orange . . . 10	**Desserts**
Hot dog 30	Coffee 10	Ice cream . . 25
Tuna 35	Tea 10	Cake 30
Beef 45		Pie 30

Choose a lunch from the menu and find what it costs.

 You have a dollar to spend for lunch. Can you find a way to spend all of it at the lunch counter?

Discussing the Ideas

How is your "mental" arithmetic? Try to find the cost of these lunches in your head.

1. Ham sandwich
   Bean soup
   Milk

2. Hamburger
   Chili
   Root beer

3. Peanut butter sandwich
   Vegetable soup
   Orange
   Pie

4. 2 Hot dogs
   Milk
   Cake

5. Beef sandwich
   Chili
   Ice cream
   Milk

6. 2 Hamburgers
   2 Milks
   Cake

Using the Ideas

1. What costs the most on the menu?

2. Which soup costs the least?

3. How much more is cake than ice cream?

4. For lunch Sue had a hot dog, vegetable soup, and a root beer. How much did her lunch cost?

5. Cindy bought a hamburger, a milk shake, and a piece of pie. How much did her lunch cost?

6. How much more is a beef sandwich than a tuna sandwich?

7. Andy had a sandwich, a soup, and a beverage. He spent the least amount that he could. How much did he spend?

8. Bill spent 77 cents. What did Bill have to drink?

★ 9. Ann had 95 cents to spend for lunch. She spent more than 85 cents. What might Ann have ordered?

★ 10. Tom did not like soup or dessert. He spent 95 cents for lunch. What might Tom have ordered if he had only one beverage?

★ 11. Jean had one soup, one beverage, and one dessert. She spent 95 cents. What did she have to drink?

● Let's explore regrouping.

Discussing the Ideas

1. Solve the equations.

A	(3 bundles) | | | | |	$35 = 30 + n$
B	(2 bundles) | | | | |	$35 = 20 + n$
C	(2 bundles) | | | | | | |	$27 = 20 + n$
D	(1 bundle) | | | | | | |	$27 = 10 + n$
E	(4 bundles) | |	$42 = 40 + n$
F	(3 bundles) | |	$42 = 30 + n$

2. Find the missing numbers.

A $48 = 40 + n$
    $48 = 30 + n$

D $45 = 40 + n$
    $45 = 30 + n$

G $42 = 40 + n$
    $42 = n + 12$

B $72 = 70 + n$
    $72 = 60 + n$

E $54 = 50 + n$
    $54 = 40 + n$

H $99 = 90 + n$
    $99 = n + 19$

C $65 = 60 + n$
    $65 = 50 + n$

F $61 = 60 + n$
    $61 = 50 + n$

I $68 = n + 8$
    $68 = 50 + n$

Using the Ideas

1. Study examples A and B.

    A To think of 52 as 40 + 12, we can write .

    B To think of 37 as 20 + 17, we can write ³²3⁷7.

    Find the matchings. Part C is matched with 4.

    C 40 + 13

    D 20 + 14

    E 60 + 12

    F 70 + 11

    G 30 + 18

    H 50 + 16

    1. ²3⁴4
    2. ⁷8¹¹1
    3. ⁵6¹⁶6
    4. ⁴5¹³3
    5. ³4¹⁸8
    6. ⁶7¹²2

2. Complete each of the following as in the examples above.

    A To think of 48 as 30 + 18, we can write ▨.

    B To think of 65 as 50 + 15, we can write ▨.

    C To think of 72 as 60 + 12, we can write ▨.

    D To think of 49 as 30 + 19, we can write ▨.

    E To think of 83 as 70 + 13, we can write ▨.

    F To think of 54 as 40 + 14, we can write ▨.

    G To think of 21 as 10 + 11, we can write ▨.

    H To think of 96 as 80 + 16, we can write ▨.

● *How is regrouping used to find differences?*

Investigating the Ideas

Look for a pattern.

43	43	43	43	43	43	43
−1	−2	−3	−4	−5	−6	−7
42	41	40				

 Can you find the rest of these differences without subtracting?

75	75	75	75	75	75	75
−32	−33	−34	−35	−36	−37	−38
43	42	41				

Discussing the Ideas

1. Explain when the number of tens changes in the sets of answers above.

2. Explain each of the steps below.

Step 1	Step 2	Step 3	Step 4
6**4** −2**6**	$\overset{5\ 14}{\cancel{6}\cancel{4}}$ −2 6	$\overset{5\ 14}{\cancel{6}\cancel{4}}$ −2 **6** **8**	$\overset{5\ 14}{\cancel{6}\cancel{4}}$ −**2** 6 **3** 8
4 − 6 = !!!	64 = 50 + 14	14 − 6 = 8	50 − 20 = 30

3. Find this difference by following the steps above. The answer is 48.

    75
  −27

4. Explain the steps for the first example. Then try the second one on your own.

$\overset{6\ 11\ 13}{\cancel{7}\cancel{2}\cancel{3}}$
−1 4 6
 5 7 7

  6 4 2
−2 5 6

Using the Ideas

1. Cover the answers and work the problems.

A	74	B	52	C	92	D	83	E	61
	−26		−35		−64		−27		−25
	48		17		28		56		36

2. Check each answer in exercise 1 by addition.

3. Find the differences.

A	43	B	33	C	72	D	54	E	81	F	68
	−16		−15		−44		−17		−23		−25

G	42	H	95	I	76	J	69	K	20	L	27
	−19		−77		−38		−62		−17		−18

M	42	N	50	O	68	P	156	Q	143	R	143
	−38		−18		−29		−82		−71		−74

S	120	T	165	U	142	V	122	W	130	X	115
	−68		−76		−65		−47		−56		−88

★ 4. Find the differences.

A	124	B	224
	−56		−56

C	132	D	432
	−95		−95

E	143	F	543
	−68		−168

G	724	H	631
	−157		−256

think 60^2

It's true our sum is 60,
And we differ by just 2.
If you want to find our names,
Some thinking you must do.
WHO ARE WE?

More practice, page A-13, Set 19

● *Can you estimate someone's weight?*

Investigating the Ideas

How close can you come to guessing someone's weight?

Name	Guess	Weight	Difference
Ann	45	52	?
Fred	60	49	?

 Can you guess the weight of each classmate in your group?

Make a chart like the one above to record your findings.

Discussing the Ideas

1. A Did you guess anyone's weight exactly?
   B Whose weight did you come closest to guessing?
   C How many times did you guess "over"?
   D How many times did you guess "under"?

2. A Did you use what you know about **your** size and weight to help you make your guesses? How?
   B How could you make better guesses?

3. A man 5 feet 8 inches tall might weigh about 150 pounds.
   A Can you find two students who together weigh this much?
   B Can you find three students who together weigh this much?

4. A What is your guess for the total weight of your class?
   B Did you miss the total weight by more than 1000 pounds?

Using the Ideas

1. Which child weighs most?
2. Who weighs least?
3. How many children weigh less than 50 pounds?
4. How many children weigh more than 60 pounds?
5. The table shows the guess and the measured weight for each student. Find how close the guess was for each student.

Name	Guess	Weight	Difference
Bobby	65	53	?
Susan	61	60	?
Tom	55	61	?
Sara	47	42	?
Alan	49	56	?
Bill	63	63	?
Jane	52	47	?
Tony	56	48	?
Joan	51	55	?
Ann	48	53	?
Rick	67	64	?

6. The boys' names are in red boxes. How much heavier is the heaviest boy than the lightest boy?

7. How much do these children weigh together?
   - A Bobby and Alan
   - B Susan and Sara
   - C Rick and Tony
   - D Ann and Joan
   - E Sara and Ann
   - F Bill and Alan

8. How much heavier is the heaviest girl than the lightest girl?

★ 9. Two children got on the scales together. The scales showed 127. Who were the children?

★ 10. How much greater is the total weight of the boys than the total weight of the girls?

think

Number thirty-one
Is slightly more than me.
I'm that much larger
Than number twenty-three.

WHO AM I?

High and Low Temperatures

Miss Smith's class decided to keep a record of high and low temperatures for days in November.

For each day, the students wrote the two temperatures on the school calendar.

NOVEMBER						
S	M	T	W	T	F	S
		1	2	3	4	5
		62 47	60 46	51 40	49 38	48 35
6	7	8	9	10	11	12
46 33	49 30	51 39	50 31	48 33	47 34	45 33
13	14	15	16	17	18	19
44 33	50 31	64 38	63 37	59 35	48 34	
20	21	22	23	24		
34 29	35 26	37 33	38 34	40 36		
27	28	29	30			
44 30	46 24	48 33	52 38			

Use the calendar above for these problems.

1. A Which Monday had the highest temperature?
   B Which Monday had the lowest temperature?

2. Water freezes at temperatures of 32 or lower. How many days are shown with freezing temperatures?

3. Which Sundays had freezing temperatures?

4. How many days are shown with temperatures above 60?

5. What was the difference between the high and the low temperature on each of these days?
   A November 2
   B November 27
   C November 1
   D November 8

6. How much more was the high temperature on November 14 than the high temperature on November 6?

7. November 15 was the warmest day of the month. November 28 was the coldest. What was the difference between the highest and the lowest temperatures in November?

110

FIGURING SCORES

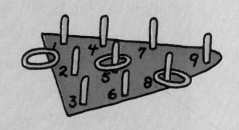

The numbers given in the chart are the children's scores for each game. For example, Dan scored 47 points in game 3.

	Ann	Bill	Carol	Dan	Ed	Fay
Game 1	28	27	62	81	60	47
Game 2	34	26	58	27	54	23
Game 3	39	43	72	47	34	52
Game 4	50	76	40	56	67	58
Game 5	63	18	43	64	29	67
Game 6	31	71	42	54	34	36

1. How many points did Fay score in games 2 and 3 together?

2. How many more points did Bill score in game 4 than in game 3?

3. In game 6, how much more was the highest score than the lowest score?

4. In game 2, how many points did Carol and Dan score together?

5. In game 5, how much less was Ed's score than Dan's?

6. Fay found the sum of her two best scores. Was this more or less than the sum of Ed's two best scores?

7. In game 1, the sum of the two highest scores was 143. In what other game was the sum of the two highest scores 143?

8. What is the difference of the highest score on the chart and the lowest score?

★ 9. The children decided that the winner should be the one who had the highest sum for his best 3 games. Who won?

More practice, page A-14, Set 20

• Let's count money.

Investigating the Ideas

Find the prices of some things you would like to buy for less than one dollar.

Item and Cost	quarters	dimes	nickles	pennies
Pen 69¢	2	1	1	4
Game 98¢	3	2	0	3

 Can you fill in a table like this by giving the **fewest** coins you could use to buy each one?

Discussing the Ideas

1. Give some other ways you could pay for the pen.

2. What are some other ways you could pay for the game?

3. What change would you get back from one dollar for each item above?

4. Give some different ways you could give someone change for a dollar.

5. How much money would you have if you had two of each coin listed in the chart?

Using the Ideas

1. Give the value of each coin collection.

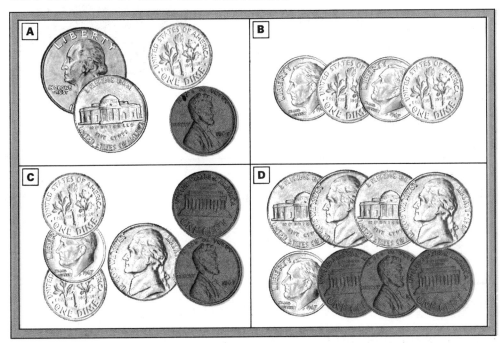

2. For each part, tell which coin collection has the greater value.
   A B or C **B** A or B **C** A or D **D** C or D

3. Give the value of each of the two coin collections together.
   A A and B **C** A and C **E** A and D
   B B and D **D** B and C **F** C and D

4. **A** How much more is A than B? **D** How much more is B than C?
   B How much more is A than C? **E** How much more is A than D?
   C How much more is C than D? **F** How much more is B than D?

5. Which pair of collections has the greater value?
   A A and B **or** C and D
   B A and C **or** B and D
   C A and B **or** A and C

★ 6. **A** Give the total value of all the coin collections together.
   B How much more would you need to have 5 dollars?

113

● *Can you add and subtract amounts of money?*

Investigating the Ideas

Choose two things you can buy at a total cost of less than 3 dollars.

 Can you fill in a chart like the one below?

Had	Bought	Bought	Spent in all	Had left
$3.00	Record $.95	Game $1.50	$2.45	$.55

Discussing the Ideas

1. Give the missing numbers.
   Example: $5.28 means 5 dollars and 28 cents.
   A $7.16 means ▥ dollars and 16 cents.
   B $4.75 means 4 dollars and ▥ cents.
   C $2.09 means ▥ dollars and 9 cents.
   D $17.34 means ▥ dollars and ▥ cents.

2. Give the missing numbers.
   Example: $4.38 is 438 cents (438¢).
   A $6.52 is ▥ ¢.   C $1.00 is ▥ ¢.   E $.50 is ▥ ¢.
   B $1.27 is ▥ ¢.   D $.76 is ▥ ¢.   F $2.86 is ▥ ¢.

3. Give the number of dollars and cents for each exercise.
   Example: 364¢ is $3.64
   A 125¢   C 506¢   E 650¢   G 1000¢   I 1250¢
   B 326¢   D 400¢   F 100¢   H 1100¢   J 1795¢

Using the Ideas

1. Find the total amounts.

   Example:
   $5.68
   2.71
   ―――
   $8.39

   A $3.27
      4.61
      ―――

   B $5.38
      1.25
      ―――

   C $7.64
      1.75
      ―――

   D $2.76
      1.85
      ―――

   E $3.79
      4.48
      ―――

   F $9.72
      8.99
      ―――

2. Find the total amounts.

   Examples:
   $1.23 and $2.45 is $3.68.
   $2.45 and $3.72 is $6.17.

   A $2.14 and $1.53 is ▩.
   B $5.21 and $2.37 is ▩.
   C $3.50 and $3.50 is ▩.

3. Find the difference in the amounts.

   Example:
   $5.27
   1.43
   ―――
   $3.84

   A $6.34
      1.52
      ―――

   B $7.65
      2.24
      ―――

   C $8.32
      4.18
      ―――

   D $9.21
      1.65
      ―――

   E $12.72
      5.28
      ―――

   F $10.64
      3.95
      ―――

Solving Story Problems

1. Debra had this much money. She bought a book for 99 cents and paid 5 cents tax.

   A How much did she have at the start?
   B How much did she spend?   C How much did she have left?

2. Craig had this much money. He bought some goldfish for $1.39. His tax was 7 cents.

   A How much did he have at the start?
   B How much did he spend?   C How much did he have left?

How sharp are your adding and subtracting skills?

1. Find the sums and differences.

A 24	B 35	C 56	D 78	E 92	F 64
+15	−12	+22	−53	−42	+23

2. Find the sums.

A 75	B 38	C 67	D 67	E 84	F 84
+16	+25	+17	+57	+92	+98

3. Find the differences.

A 52	B 52	C 73	D 73	E 126	F 126
−16	−36	−18	−38	−53	−59

4. Find the sums and differences.

A 75	B 67	C 56	D 70	E 78	F 156
+83	+54	−9	−26	+65	−88

G 132	H 324
−54	+247

I 384	J 432
+247	−154

5. Find the sums.

A 21	B 34
25	23
+23	+18

★ **6.** Find the differences.

A 602	B 500
−24	−34

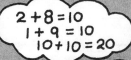

think

2 + 8 = 10
1 + 9 = 10
10 + 10 = 20

Find each sum quickly without pencil and paper.

1. 1 + 5 + 9
2. 1 + 2 + 5 + 8 + 9
3. 1 + 2 + 3 + 5 + 7 + 8 + 9
4. 1 + 50 + 99
5. 1 + 2 + 50 + 98 + 99
6. 1 + 2 + 3 + 50 + 97 + 98 + 99

Solving Short Story Problems

1. 45 cents for a sandwich.
15 cents for milk.
How much for both?

2. Had a dollar.
Spent 49 cents.
How much left?

3. Weighed 58 pounds.
Gained 14 pounds.
Weigh how much now?

4. 52 marbles.
17 are taken away.
How many are left?

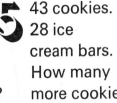

5. 43 cookies.
28 ice cream bars.
How many more cookies than ice cream bars?

6. 84 cents for a game. 56 cents for a puzzle.
How much less for the puzzle than for the game?

7. 14 apples in a sack.
More are put in.
32 apples in the sack now.
How many were put in?

8. Lunch period: 45 minutes.
Recess: 25 minutes.
How much longer for lunch?

9. Room A: 32 children.
Room B: 27 children.
How many more in Room A than in B? How many less in Room B than in A?

10. 65 cars. 19 are black. How many are not black?

11. 60 eggs.
One dozen are broken.
How many are not broken?

12. 45 doughnuts.
27 of them are eaten.
How many are left?

★ **13.** 25 children.
34 lollipops.
Each child gets at least one lollipop. How many children can have two?

More practice, page A-16, Set 22

Reviewing the Ideas

1. Find the sums and the differences.

   A. 23 + 51
   B. 78 − 24
   C. 56 + 13
   D. 60 + 16
   E. 57 − 36
   F. 82 − 32

   G. 27 + 54
   H. 62 − 24
   I. 48 + 36
   J. 75 + 15
   K. 43 − 15
   L. 51 − 27

   M. 67 + 27
   N. 86 − 67
   O. 70 − 26
   P. 38 + 57
   Q. 39 + 48
   R. 81 − 44

2. Find the sums.

   A. 23 + 16 + 50
   B. 31 + 26 + 12
   C. 42 + 14 + 25
   D. 30 + 50 + 20
   E. 32 + 57 + 26
   F. 38 + 24 + 10

3. Find the sums and differences.

   A. 78 + 49
   B. 120 − 57
   C. 95 + 36
   D. 137 − 68
   E. 87 + 76
   F. 153 − 75

4. Find the differences.

   A. 326 − 114
   B. 9782 − 1542
   C. 3641 − 1216
   D. 4372 − 144

★ 5. Find the differences.

   A. 1682 − 427
   B. 8002 − 643

think

Diane has 48 cents.

1. What coins does she have if she has 9 coins?
2. What is the fewest number of coins she could have?
3. What is the fewest number of coins she could have and still have at least one of each coin pictured?

6. Find the amounts.

   A $1.25   B $3.84   C $1.84   D $6.40   E $7.04   F $5.00
      +2.41   −1.22   +2.09   −1.25   +1.88   −4.98

7. In these exercises, no numbers are given. You should decide whether you would add or subtract if numbers were given. Answer **A** if you would **add** to answer the question. Answer **S** if you would **subtract**. Think carefully before giving your answer.

   A Joe had ||||| marbles. He gave ||||| to Tom. How many did he have left?

   B Jane counted ||||| cars on one train and ||||| cars on another. How many did she count in all?

   C Sue had ||||| cents. She spent ||||| cents for an apple. How much did she have left?

   D Tom is ||||| years old, and Jim is |||||. How much younger is Jim than Tom?

   E Susan spent ||||| cents for a sandwich and ||||| cents for milk. How much did she spend?

   F In Jane's class there are ||||| children. There are ||||| boys in the class. How many girls are in Jane's class?

   G Tim checked ||||| books out of the library. Later he returned ||||| of the books. How many did he still have?

   H Beth and Sara have ||||| dolls in all. ||||| of the dolls are Sara's. How many of the dolls are Beth's?

   I Jack read ||||| pages in his book. He has ||||| pages yet to read. How many pages are in the book?

   J Ann said, "||||| years ago, I was ||||| years old." How old is Ann now?

Keeping in Touch with — Addition, Subtraction, Place value, Inequalities

1. Four of these statements are false. Which ones are they?

   A $15 - 5 = 5$
   B $26 + 7 > 30$
   C $34 - 8 > 30$
   D $39{,}427 > 38{,}998$
   E $48 - 9 < 40$
   F $48 - 19 > 30$
   G $67 + 8 = 70 + 5$
   H $48{,}265 = 48{,}165 + 1000$
   I $528 = 500 + 20 + 8$

2. Write the correct numeral for each ▨.

   A For 6 tens and 2, we write ▨.
   B For 8 tens and 7, we write ▨.
   C For 2 tens and 5, we write ▨.
   D For 5 tens and 2, we write ▨.
   E For 7 tens and 0, we write ▨.
   F For 1 ten and 0, we write ▨.

3. Solve the equations.

   A $6 + 5 = n$
   B $4 + 3 = n$
   C $8 + 7 = n$
   D $9 + 2 = n$
   E $6 + 4 = n$
   F $8 + 8 = n$
   G $n + 6 = 8$
   H $n + 4 = 11$
   I $n + 9 = 10$
   J $n + 8 = 12$
   K $n + 6 = 6$
   L $n + 5 = 10$
   M $14 - 5 = n$
   N $9 - 6 = n$
   O $12 - 5 = n$
   P $11 - 4 = n$
   Q $15 - 7 = n$
   R $18 - 9 = n$
   S $n + 6 = 8$
   T $4 + n = 9$
   U $6 + 8 = n$
   V $n - 2 = 4$
   W $9 - n = 4$
   X $15 - 5 = n$

4. Copy each column and complete the counting.

   A 6, 7, 8, ▨, ▨, ▨, ▨
   B 16, 17, 18, ▨, ▨, ▨, 22
   C 86, 87, 88, 89, ▨, ▨, ▨
   D 96, 97, 98, ▨, ▨, ▨, ▨
   E 436, 437, 438, 439, ▨, ▨, ▨

5. Solve each equation.

   A $9 + 1 = n$ B $99 + 1 = n$ C $999 + 1 = n$ D $9999 + 1 = n$

6. Give the correct sign $<$ or $>$ for each ⬤.

A 26 ⬤ 28	I 238 ⬤ 228	Q 672 ⬤ 658
B 94 ⬤ 90	J 654 ⬤ 664	R 180 ⬤ 159
C 87 ⬤ 77	K 275 ⬤ 272	S 338 ⬤ 242
D 59 ⬤ 99	L 542 ⬤ 546	T 665 ⬤ 571
E 127 ⬤ 147	M 556 ⬤ 561	U 862 ⬤ 955
F 152 ⬤ 122	N 474 ⬤ 439	V 347 ⬤ 274
G 219 ⬤ 209	O 621 ⬤ 612	W 631 ⬤ 713
H 257 ⬤ 277	P 389 ⬤ 398	X 865 ⬤ 568

7. Copy the problems and give the missing digits.

   A   1▇   B   26
      − 6   + ▇
      ────   ────
       11   34

   C   ▇6   D   6▇
      +19   + ▇7
      ────   ────
       4▇   91

   E   72   F   ▇▇ ▇▇
      −3▇   −47
      ────   ────
       ▇8   16

   G   43▇   H   48▇
      +2▇3   +▇54
      ────   ────
      ▇98   9▇3

think

Nan picked out 5 white shells. She put some blue paint on 3 of them. Then she could see white on only 3 of the 5 shells. On how many could she see both blue and white?

You are invited to explore

ACTIVITY CARD 5
Page 311

6 Multiplication

● *What is multiplication?*

Investigating the Ideas

Get some paper cups, counters, and a collection box.

Number of cups	Number in each cup	Total number in product box	Multiplication equation
5	4	20	5 × 4 = 20

 Can you use different numbers of cups and counters and write 4 more multiplication equations?

Discussing the Ideas

1. Give the missing numbers and explain how to write a multiplication equation.

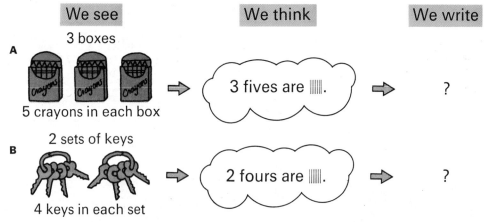

2. Make up an example like those above.

Using the Ideas

1. A How many words?
    B How many letters in each word?
    C How many letters in all?
    D Solve: $5 \times 3 = n$

    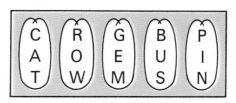

2. A How many pairs of shoes?
    B How many in each pair?
    C How many shoes in all?
    D Solve: $4 \times 2 = n$

3. A How many ants?
    B How many legs on each ant?
    C How many legs in all?
    D Solve: $3 \times 6 = n$

4. A How many sets?
    B How many dots in each set?
    C How many dots in all?
    D Solve: $4 \times 3 = n$

    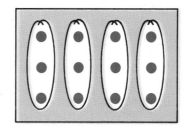

5. A How many nickels?
    B How many cents for each nickel?
    C How many cents in all?
    D Solve: $5 \times 5 = n$

6. Solve the equation.

    $3 \times 4 = n$

7. Write a multiplication equation.

    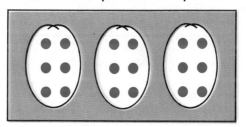

● *Can the number line help you think about multiplication?*

Investigating the Ideas

You can use your strips and a centimeter ruler to help you think about multiplication.

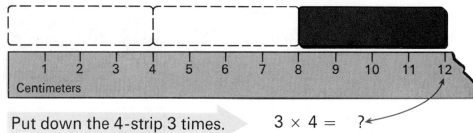

Put down the 4-strip 3 times. $3 \times 4 = $ ___?___

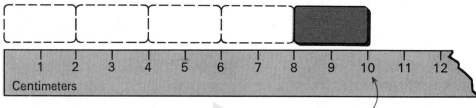

Put down the 2-strip 5 times. $5 \times 2 = $ ___?___

 Can you use your strips and a ruler and write at least six more multiplication equations?

Discussing the Ideas

1. What multiplication equation does the picture suggest?

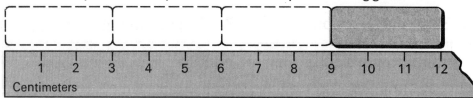

2. What multiplication fact does the number line show?

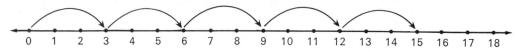

Using the Ideas

1. Write a multiplication equation for each picture.

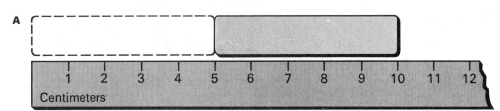

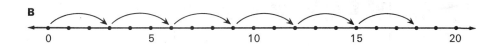

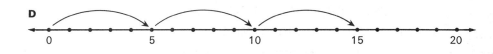

2. You can use your centimeter ruler to draw number lines.

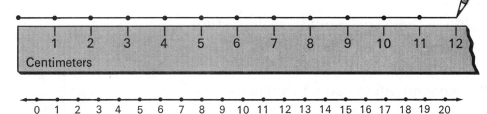

Draw 3 number lines. Use them to show these multiplications.

   A 2×7 　　　　B 4×3 　　　　C 4×4

★ 3. Find the missing numbers.

   A 2, 4, 6, ▮, ▮, ▮, ▮, 16 　　　E 6, 12, 18, ▮, ▮, ▮, ▮, 48
   B 3, 6, 9, ▮, ▮, ▮, ▮, 24 　　　F 7, 14, 21, ▮, ▮, ▮, ▮, 56
   C 4, 8, 12, ▮, ▮, ▮, ▮, 32 　　G 8, 16, 24, ▮, ▮, ▮, ▮, 64
   D 5, 10, 15, ▮, ▮, ▮, ▮, 40 　　H 9, 18, 27, ▮, ▮, ▮, ▮, 72

● *Let's think about multiplication in another way.*

Investigating the Ideas

Cut five thin red strips and five thin blue strips.
 Now make a table like the one below.

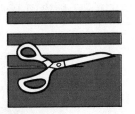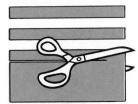

Number of red strips	Number of blue strips	Number of crosses	Multiplication equation
3	4	12	3 × 4 = 12
5	3	?	5 × 3 = ?

 Can you use your strips to help you write other multiplication equations?

Discussing the Ideas

1. Show how you would use your strips to solve this equation.

    4 × 2 = **n**

2. What equation can you solve by laying the strips down like this? Show how to do this.

3. You could draw lines instead of using your strips. What equation can you solve by counting crosses on these lines?

    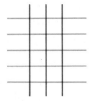

4. What are the largest numbers you could multiply using your strips?

Using the Ideas

1. Write and solve a multiplication equation for each picture.

   A   **B**   **C**

   Parts of the lines have been erased in exercises D, E, and F, but you can still see the crosses.

   D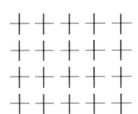

   E
   ```
   + + +
   + + +
   + + +
   + + +
   ```

   F
   ```
   + + +
   + + +
   + + +
   + + +
   + + +
   + + +
   + + +
   ```

   Only the dots where the lines cross are left in exercises G, H, and I, but you can still tell how many lines and how many crosses.

   G    **H**

   I
   ```
   • • • • • • •
   • • • • • • •
   • • • • • • •
   • • • • • • •
   • • • • • • •
   ```

2. Use strips, lines, or dots and solve these equations.

   A $2 \times 6 = ?$ **B** $4 \times 5 = ?$ **C** $3 \times 7 = ?$ **D** $8 \times 3 = ?$

• *How are addition and multiplication related?*

Investigating the Ideas

Find the sum for each picture.

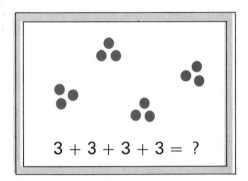

$3 + 3 + 3 + 3 = ?$

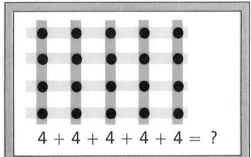

$4 + 4 + 4 + 4 + 4 = ?$

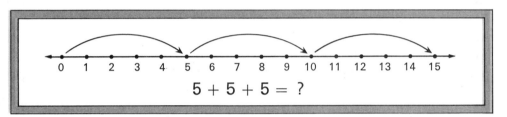

$5 + 5 + 5 = ?$

 Can you write a multiplication equation for each picture?

Discussing the Ideas

1. Carol said, "Multiplication is just a short way to do addition." What do you think she meant?

2. A Can you make up an example using sets to show $6 + 6 + 6$? What multiplication equation goes with this?

   B Can you make up an example using dots to show $4 + 4$? What multiplication equation goes with this?

   C Can you make up an example using the number line to show $3 + 3 + 3 + 3 + 3$? What multiplication equation goes with this?

Using the Ideas

1. A How many tricycles?
   B How many wheels on each tricycle?
   C How many wheels in all?
   D Write an addition equation about the sets.
   E Write a multiplication equation about the sets.

2. A How many bowls?
   B How many fish in each bowl?
   C How many fish in all?
   D Write one addition and one multiplication equation about the sets.

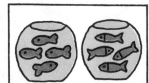

3. Solve the equations.
   A $4 + 4 + 4 = n$
   $3 \times 4 = n$

   B $6 + 6 + 6 = n$
   $3 \times 6 = n$

   C $3 + 3 + 3 + 3 + 3 = n$
   $5 \times 3 = n$

   D $2 + 2 + 2 + 2 = n$
   $4 \times 2 = n$

4. A Ken made a dart board. The first time he threw 6 darts, the board looked like this. Nancy was scorekeeper. She wrote $2 + 2 + 2 + 2 + 2 = 10$. Ken found his score by multiplication. Give the multiplication problem Ken worked.
   B In one game Bob threw all 6 darts in the 4 ring. What was his score?
   ★ C Jane threw 3 darts. Her score was 14. Where did the darts land?
   ★ D John threw 3 darts. His score was 12. Where could his darts have landed?

COIN COLLECTION PROBLEMS
A Penny Collection

1. Some coin collectors keep their coins in a folder. In the folder page above, how many pennies are in each row? How many rows are there? How many pennies in all?

2. In the folder page above, how many pennies are in each column? How many columns are there? How many pennies in all?

3. Jim collects pennies. He made wrappers for some of them. He put 5 pennies in each roll. He put 4 rolls of pennies in a box. How many pennies were in the box?

4. Coin collectors often pay more than 1 cent for certain pennies, to complete their collections. Jim's book lists the value of a certain old penny as 5 cents. Jim has three of these pennies. How much are they worth?

5. Jim had 3 rows of Indian Head pennies with 6 in each row. How many Indian Head pennies did he have?

★ 6. Suppose a certain penny is worth 5 cents and an older penny is worth 10 cents. Which is worth more, five of the pennies worth 5 cents or three of the pennies worth 10 cents?

A Dime Collection

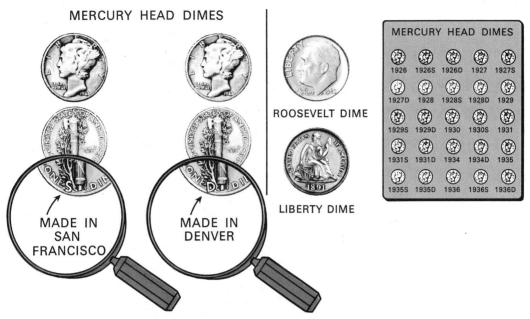

MERCURY HEAD DIMES

ROOSEVELT DIME

LIBERTY DIME

MADE IN SAN FRANCISCO

MADE IN DENVER

1. In the folder page, how many dimes are in each row? How many rows are there? How many dimes in all?

2. There are 3 pages in each folder. Bill has 7 folders. How many pages does Bill have in his folders?

3. Bill put 10 dimes in a roll. How many cents is the roll worth?

4. Bill put 4 rolls in a box. How many dimes did Bill put in the box?

5. Ted had 6 rows of Liberty dimes with 3 dimes in each row. How many Liberty dimes did he have?

★ 6. 15,840,000 Mercury dimes were made in San Francisco in 1935. 159,130,000 Mercury dimes were made in Philadelphia in 1945. Which of these dimes do you think Jim is most likely to find?

★ 7. One page has spaces for 7 rows of Roosevelt dimes with 4 dimes in each row. There are only 7 empty spaces. How many Roosevelt dimes are already on the page?

● *Let's explore factors and products.*

Investigating the Ideas

Make cards like these.

Symbol cards   Factor cards   Product cards

Here is one multiplication equation you can make with your cards.

 How many more multiplication equations can you make? | Record each one.

Discussing the Ideas

The numbers we multiply are called factors of the product.
The answer in multiplication is called the product of the factors.

1. Suppose the product is 24. One factor is 6. The other factor is 4. How would you write this as in A? as in B?

2. How are factors and products like addends and sums?

Using the Ideas

1. **A** In 3 × 5 = 15, the number 15 is the product of what factors?
   B In 4 × 3 = 12, what is the number 12 called?
   C In 6 × 7 = 42, the number 42 is the product of what factors?
   D In 8 × 9 = 72, what is the number 72 called?
   E In 3 × 5 = 15, the numbers 3 and 5 are factors of what product?
   F In 4 × 3 = 12, what are the numbers 4 and 3 called?

2. Write a multiplication equation for each exercise. Write an **f** over each factor and a **p** over the product as in the example:
   $\overset{f}{3} \times \overset{f}{4} = \overset{p}{12}$

   A

   B 4 + 4 + 4 + 4 + 4 = 20

   C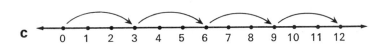

3. What is the product when
   A 2 and 3 are the factors?
   B 3 and 4 are the factors?
   C 2 and 7 are the factors?
   D 9 and 2 are the factors?
   E 3 and 7 are the factors?
   F 6 and 4 are the factors?

think

Alan and Bob together weigh 90 pounds.
Alan, Bob, and Jim together weigh 148 pounds.

1. How much does Jim weigh?

Alan and Jim together weigh 105 pounds.

2. How much does Alan weigh?

3. How much does Bob weigh?

Short Picture Problems

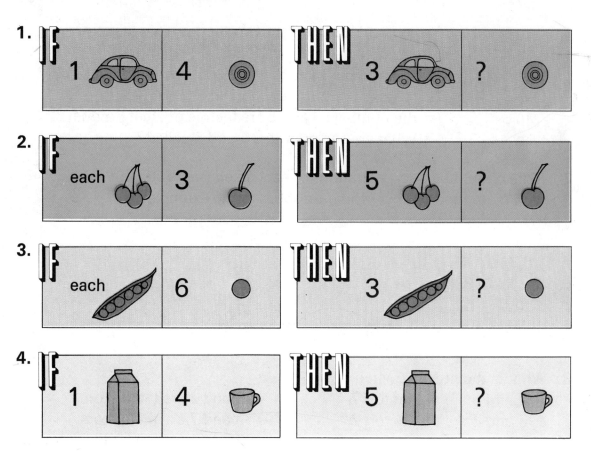

Short Stories

1. 3 engines on each jet.
5 jets.
How many engines?

2. 4 chairs in each row.
3 rows.
How many chairs?

3. 5 puppies.
4 legs per puppy.
How many legs?

4. 6 legs on an insect.
3 insects.
How many legs?

5. 8 legs on a spider.
2 spiders.
How many legs?

Keeping in Touch with   Addition   Inequalities
  Subtraction   Basic principles
  Fractions   Measurement

1. For each pair, write the larger number on your paper.
   A 3764; 4764   B 67,289; 67,290

2. A What is the area of the rectangle?
   B What is the area of $\frac{1}{2}$ of the rectangle?
   C What is the area of $\frac{1}{4}$ of it?

3. Solve the equations.
   A $295 = 200 + n + 5$ B $6285 = 6000 + n + 80 + 5$

4. Write 2 addition and 2 subtraction equations for each exercise.

   A    B    C

5. Solve the equations.
   A $6 + 7 = n$ B $7 + 6 = n$ C $8 + 4 = n$ D $4 + 8 = n$

6. Solve the equations.
   A $(4 + 3) + 6 = n$ C $(5 + 2) + 4 = n$
   B $4 + (3 + 6) = n$ D $5 + (2 + 4) = n$

7. Find the sums and differences.

   | A 34 | B 45 | C 74 | D 93 | E 26 | F 32 |
   | +27 | +38 | −26 | −47 | +49 | −18 |

   | G 85 | H 123 | I 48 | J 97 | K 84 | L 147 |
   | +69 | −64 | +75 | −69 | +66 | −75 |

    You are invited to explore   ACTIVITY CARD 6 Page 312

● *Let's explore multiplying by 0 and 1.*

Investigating the Ideas

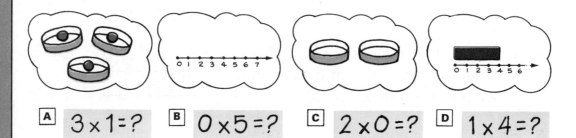

A $3 \times 1 = ?$ B $0 \times 5 = ?$ C $2 \times 0 = ?$ D $1 \times 4 = ?$

 How many of these products can you give correctly? Check your answers with a classmate.

Discussing the Ideas

1. Think of cups and counters.
   A How does this show 4×1?
   B How would you show 1×4?
   C How would you show 3×0?

2. Think of strips and a number line.
   A How does this show 1×5?
   B How would you show 5×1?

   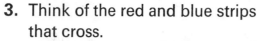

3. Think of the red and blue strips that cross.
   A How does this show 4×1?
   B How would you show 4×0?
   C How would you show 0×4?

   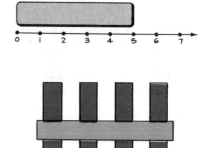

4. Can you give a rule for multiplying by 0? by 1?

Using the Ideas

1. Write a multiplication equation for the figure.

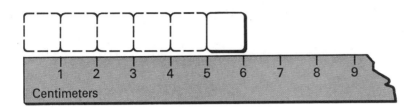

2. Complete the statements.
   A The product of any number and **1** is __?__.
   B The product of any number and **0** is __?__.

3. Find the products.
   A $9 \times 0 = n$
   B $0 \times 9 = n$
   C $7 \times 1 = n$
   D $1 \times 7 = n$
   E $12 \times 0 = n$
   F $1 \times 12 = n$
   G $35 \times 1 = n$
   H $0 \times 35 = n$
   I $97 \times 0 = n$

4. Solve the equations.
   A $8 \times n = 0$
   B $n \times 7 = 7$
   C $0 \times 6 = n$
   D $19 \times n = 0$
   E $1 \times 56 = n$
   F $74 \times n = 74$
   G $n \times 62 = 62$
   H $58 \times n = 0$
   I $0 \times 254 = n$

★ 5. Some of these equations have just **one** solution, some have **many** solutions, and some have **no** solution. Answer: **one**, **many**, or **none**.
   A $n \times 6 = 0$
   B $0 \times n = 6$
   C $0 \times n = 0$
   D $6 \times 0 = n$
   E $n \times 0 = 0$
   F $n \times 0 = 8$

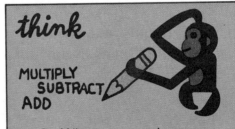

think

MULTIPLY
SUBTRACT
ADD

1. What two numbers have a product of 20 and a difference of 19?

2. Find two numbers so that their product is less than their sum.

● *Let's explore the order principle.*

Investigating the Ideas

Here are some special pairs of "matching" trains.

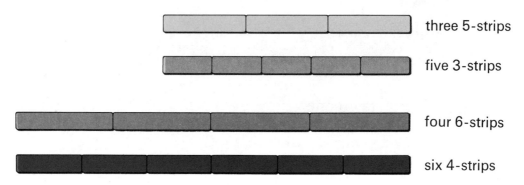

three 5-strips

five 3-strips

four 6-strips

six 4-strips

 Can you make some more special pairs of trains like these? | Record each pair you make.

Discussing the Ideas

1. A Do these two trains match?
   B Can you write an equation about this?

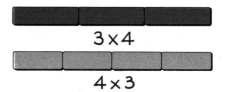

3 × 4

4 × 3

2. Can you write an equation for each pair of trains in the Investigation?

3. Think of cups and counters.
   A How does this show that $3 \times 2 = 2 \times 3$?
   B How would you use this idea to show that $5 \times 4 = 4 \times 5$?

3 cups, 2 in each   3 × 2

2 cups, 3 in each   2 × 3

4. When we change the order of the __?__, we get the same __?__.

Using the Ideas

1. Answer the questions in the Short Stories for each part.
   Then write a multiplication equation for the exercise.

   A Three 6-cent stamps cost how much?
      (Answer: 18¢)

      Six 3-cent stamps cost how much?
      (Answer: 18¢)
      (Equation: $3 \times 6 = 6 \times 3$)

   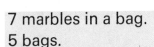

   C 5 marbles in a bag.
      7 bags.
      How many marbles?

      7 marbles in a bag.
      5 bags.
      How many marbles?

   B 4 steps.
      Each 2 feet long.
      How far?

      2 steps.
      Each 4 feet long.
      How far?

   D 2 boys.
      3 arrows each.
      How many arrows?

      3 boys.
      2 arrows each.
      How many arrows?

   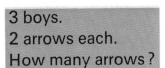

2. We can think about 15 dots in two different ways.
   Solve the equations.

   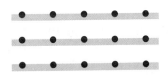

   $5 \times 3 = n$ $3 \times 5 = n$

3. Find the products.
   Use the table.

   A 92×65 D 71×17
   B 24×84 E 82×34
   C 47×38 F 27×56

$56 \times 27 = 1512$	$65 \times 92 = 5980$
$34 \times 82 = 2788$	$17 \times 71 = 1207$
$38 \times 47 = 1786$	$84 \times 24 = 2016$

● *Let's explore rearranging factors.*

Investigating the Ideas

Make three slips of paper like these.
Then turn them over and mix them up.

Pick any **two** slips and multiply the numbers on them. Then multiply by the number on the other slip.

 If you do this five times, will you get the same final product each time?

Discussing the Ideas

With the factors 2, 3, and 4

| A | we could multiply these first. | B | we could multiply these first. | C | we could multiply these first. |

$2 \times 3 \times 4$ or $2 \times 3 \times 4$ or $2 \times 3 \times 4$
$(2 \times 3) \times 4$ $2 \times (3 \times 4)$ $(2 \times 4) \times 3$

1. Answer these questions for A, B, and C. Which two factors are grouped together? What is their product? What is the final product?

2. If we leave the order of three factors the same, we can state this principle about **grouping**:

   When we multiply, we can change the **grouping** and get the same product.

In which example above did we change both **order** and **grouping**?

Using the Ideas

1. Find each product. Use the groupings shown. In each part, check to see that the two different groupings give the same product.
   - A $(5 \times 1) \times 6$
     $5 \times (1 \times 6)$
   - B $(2 \times 2) \times 4$
     $2 \times (2 \times 4)$
   - C $(4 \times 2) \times 5$
     $4 \times (2 \times 5)$

2. Find each product. Choose the grouping that is most helpful. Do not change the order.
   - A $7 \times 5 \times 2$
   - B $8 \times 10 \times 10$
   - C $593 \times 497 \times 0$

3. Solve the equations.
   - A $(3 \times 7) \times 5 = n \times (7 \times 5)$
   - B $17 \times (4 \times 29) = (n \times 4) \times 29$

4. Find the products.
   - A Since $(4 \times 6) \times 3 = 72$, we know that $4 \times (6 \times 3) = n$.
   - B Since $(5 \times 7) \times 2 = 70$, we know that $5 \times (7 \times 2) = n$.

5. Find the products. Arrange the factors any way you choose.
   - A $5 \times 8 \times 2$
   - B $2 \times 9 \times 5$
   - C $4 \times 2 \times 1$
   - D $989 \times 7 \times 0$

6. Find the products.
   - A Since $3 \times 4 \times 5 = 60$,
     we know that $5 \times 3 \times 4 = n$.
   - B Since $8 \times 7 \times 6 = 336$,
     we know that $7 \times 8 \times 6 = n$.

7. Find the products. Use the table on the right.
   - A $3 \times 27 \times 6$
   - B $5 \times 8 \times 13$
   - C $4 \times 65 \times 9$
   - D $17 \times 4 \times 6$
   - E $4 \times 27 \times 5$
   - F $65 \times 8 \times 4$
   - G $3 \times 7 \times 17$
   - H $13 \times 9 \times 8$

$5 \times 27 \times 4 = 540$
$17 \times 6 \times 4 = 408$
$8 \times 13 \times 9 = 936$
$6 \times 3 \times 27 = 486$
$17 \times 7 \times 3 = 357$
$8 \times 5 \times 13 = 520$
$65 \times 9 \times 4 = 2340$
$8 \times 4 \times 65 = 2080$

● *Let's explore the multiplication-addition principle.*

Investigating the Ideas

Cut three 8 by 4 rectangles from graph paper. Color each one a different color.

One way to think about these 8 fours is shown by this cut.

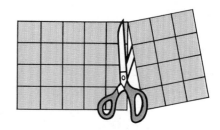

5 fours and 3 fours

 Can you make different cuts in your rectangles to show other ways to think about 8 fours? | Record your results as shown above.

Discussing the Ideas

1. To help you understand the multiplication-addition principle, think of "breaking apart" a factor before you multiply. Find the missing numbers.

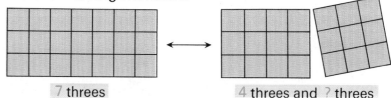

7 threes   4 threes and ? threes

(7 × 3)   equals   (4 × 3) + (? × 3)

2. These are other examples of the multiplication-addition principle. Give the missing numbers.
   A 8 fives equals 6 fives and __?__ fives.
   B 6 eights equals __?__ eights and 2 eights.
   C 7 sixes equals 5 sixes and __?__ sixes.

Using the Ideas

1. Solve the equations.

   We see: [grid of 5×3] ⟷ [grid of 3×3]) [grid of 2×3]

   (We think): 5 threes ⟷ 3 threes and 2 threes

   We write: $5 \times 3 = n$ ⟷ $(3 \times 3) + (2 \times 3) = n$

2. Give the missing number of twos.

   A 7 sets of 2 → For 7 sets of two, we can think 6 twos and ▥ twos.

   B 7 sets of 2 → For 7 sets of two, we can think 5 twos and ▥ twos.

   C 7 sets of 2 → For 7 sets of two, we can think 4 twos and ▥ twos.

3. Solve the equations.
   A $7 \times 2 = (6 \times 2) + (n \times 2)$ C $7 \times 2 = (4 \times 2) + (n \times 2)$
   B $7 \times 2 = (5 \times 2) + (n \times 2)$ D $7 \times 2 = (3 \times 2) + (n \times 2)$

4. Give the missing number. Then solve the equation.

   A For 6 sets of three, we can think 4 threes and ▥ threes.
      $6 \times 3 = (4 \times 3) + (n \times 3)$

   B For 5 sets of four, we can think 2 fours and ▥ fours.
      $5 \times 4 = (2 \times 4) + (n \times 4)$

   C For 8 sets of two, we can think 4 twos and ▥ twos.
      $8 \times 2 = (4 \times 2) + (n \times 2)$

   D For 4 sets of six, we can think 3 sixes and ▥ sixes.
      $4 \times 6 = (3 \times 6) + (n \times 6)$

● Can the multiplication-addition principle help you find products?

Investigating the Ideas

Suppose you don't know 7 × 6.

 Can you show a way to find this product by using other products that you do know?

Discussing the Ideas

1. Here is the way Eric thought about using the multiplication-addition principle to find 7 × 6.

   When you want to multiply by a number   *you can multiply by part of it*   *and then by the rest of it*

   7 × 6   =   (4 × 6)   +   (3 × 6)

   Explain how to use the same idea to find 8 × 5.

2. Diane is thinking about the multiplication-addition principle. Explain Diane's thinking.

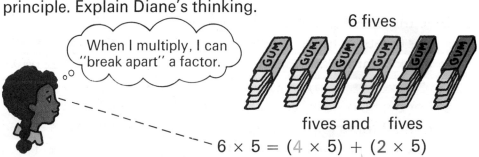

When I multiply, I can "break apart" a factor.

6 fives

fives and fives

6 × 5 = (4 × 5) + (2 × 5)

Using the Ideas

1. Solve the equations.

   A   7 fours → (? fours and 3 fours)

   $7 \times 4 = (n \times 4) + (3 \times 4)$

   B   7 fours → (5 fours and ? fours)

   $7 \times 4 = (5 \times 4) + (n \times 4)$

   C   7 fours → (6 fours and ? fours)

   $7 \times 4 = (6 \times 4) + (n \times 4)$

   D   8 fives → (? fives and 6 fives)

   $8 \times 5 = (n \times 5) + (6 \times 5)$

   E   8 fives → (7 fives and ? fives)

   $8 \times 5 = (7 \times 5) + (n \times 5)$

think

Tom is 4 years older than his younger sister and 6 years older than his younger brother. The sum of their ages is 26. How old is each child?

2. Find the products.

   A $2 \times 2 = 4$
       $3 \times 2 = 6$ $5 \times 2 = n$

   B $2 \times 3 = 6$
       $3 \times 3 = 9$ $5 \times 3 = n$

   C $6 \times 5 = 30$
       $2 \times 5 = 10$ $8 \times 5 = n$

   D $3 \times 2 = 6$
       $4 \times 2 = 8$ $7 \times 2 = n$

   E $3 \times 3 = 9$
       $4 \times 3 = 12$ $7 \times 3 = n$

   F $3 \times 7 = 21$
       $5 \times 7 = 35$ $8 \times 7 = n$

3. Find the products.

   A   Since $5 \times 8 = 40$, we know that $6 \times 8 = n$.

   B   Since $5 \times 9 = 45$, we know that $6 \times 9 = n$.

   C   Since $6 \times 8 = 48$, we know that $7 \times 8 = n$.

   D   Since $7 \times 6 = 42$, we know that $8 \times 6 = n$.

More practice, page A-17, Set 23.

● *Let's look at some of the easier multiplication facts.*

Discussing the Ideas

1. "0" and "1" facts
   A. What do you know about multiplying by **0** and **1** that will help you fill in the 0 and 1 rows in a multiplication table?
   B. Where are the "0" and "1" columns in the table? What makes it easy to fill in these columns?

2. "2" facts
   How can the "1" facts help you find the "2" facts? The picture may help.

   2 sixes are 1 six and 1 six.
   $2 \times 6 =$ 6 + 6

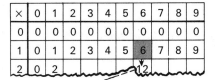

3. Give the products for A through H in this table.

4. A. If you know the product for B, it is easy to find the product for I. Why?
   B. Give the products for I through O.

Using the Ideas

1. Study the picture to see how the "1" and "2" facts can help you find the "3" facts.

   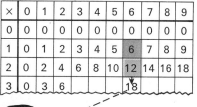

   3 sixes are 2 sixes and 1 six.
   3 × 6 = 12 + 6

   A What are 2 sevens and 1 seven?
   B What is 3 × 7?

2. Give the products A through G in the table.

3. Use the order principle and the facts in the "3" row to quickly give the products for the "3" column (H through M).
   Example: 3 × 4 = 12 (B),
   so 4 × 3 = 12 (H).

4. Find the products.

   A 6 × 1   C 4 × 3   E 3 × 8   G 5 × 3   I 7 × 3
   B 2 × 8   D 3 × 0   F 0 × 9   H 1 × 8   J 0 × 5

5. Find the products.

   A 6   B 1   C 4
      ×2   ×7   ×0

   D 4   E 3   F 8
      ×3   ×3   ×2

   G 3   H 2   I 9
      ×6   ×7   ×3

6. Copy and give the missing numbers.

   A 2, 4, 6, 8, ▨, ▨, ▨, 16, ▨
   B 3, 6, 9, ▨, ▨, 18, ▨, ▨, 27

7. Give the missing products.

×	5
1	A
2	B
3	C

×	8
1	D
2	E
3	F

×	9
1	G
2	H
3	I

More practice, page A-17, Set 24

● **What are the facts when 4 and 5 are factors?**

Discussing the Ideas

1. If you have learned the "0," "1," "2," and "3" facts, how many more multiplication facts in the table do you have left to learn?

2. A Study the small table.

   How can the "1" and "3" facts help you find the "4" facts? How can the "2" facts help you find the "4" facts?

   B Find facts for **A** through **F**. Then find facts for **G** through **K**.

3. A Study the small table. How can "2" and "3" facts help you find the "5" facts?

   B Find the facts for **A** through **E**. Then find the facts for **F** through **I**.

Using the Ideas

1. Copy and complete.

   A 4, 8, 12, ▓, ▓, 24, ▓, ▓, 36
   B 5, 10, ▓, ▓, ▓, ▓, 35, ▓, ▓

2. Copy and complete each table.

   A

×	6
1	
3	
4	

   B

×	8
1	
2	
4	

   C

×	9
1	
3	
4	

   D

×	7
1	
4	
5	

   E

×	8
2	
3	
5	

   F

×	9
2	
4	
5	

3. Find the products.

A 1×4	F 6×4	K 4×8	P 4×5	U 9×5
B 2×4	G 7×4	L 4×6	Q 5×5	V 5×4
C 3×4	H 8×4	M 1×5	R 6×5	W 5×6
D 4×4	I 9×4	N 2×5	S 7×5	X 5×5
E 5×4	J 4×7	O 3×5	T 8×5	Y 5×9

4. If you know the facts up through the "5" facts, how many facts do you have left to learn?

★ 5. Use the "5" and the "4" facts to find these "9" facts.

   A 9×6 C 9×7
   B 9×8

think

Add the numbers
 one, two, three.
Now find their product too.
I'm the answer either way,
Whichever one you do.

WHO AM I?

More practice, page A-18, Set 25.

Function Machine Problems

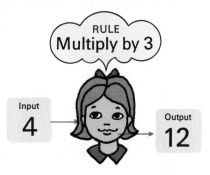

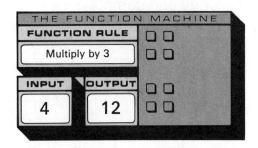

Think about the function machine and give the missing number or rule.

1. Function Rule: Multiply by 3

	Input	Output
	2	6
	9	27
A	8	24
B	5	15
C	7	21

2. Function Rule: Multiply by 4

	Input	Output
	2	8
A	4	
B	9	
C	6	
D	8	

3. Function Rule: Multiply by 5

	Input	Output
A	3	
B	7	
	9	45
C	6	
D	8	

4. Function Rule

	Input	Output
A		
	2	10
	3	15
	7	35
B	6	
C		20

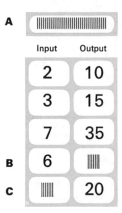

★ 5. Function Rule: Multiply by 2

	Input	Output
	2	4
A	5	
B		8
C		14
D		12

★ 6. Function Rule

	Input	Output
A		
	2	4
	3	9
	5	25
B	4	
C	1	

Multiplication Practice

1. Find the products.

"0" facts
- A 6×0
- B 0×3
- C 4×0
- D 9×0
- E 3×0
- F 0×5
- G 0×1
- H 0×0
- I 7×0
- J 5×0
- K 1×0
- L 0×8

"1" facts
- A 1×4
- B 3×1
- C 6×1
- D 1×8
- E 7×1
- F 4×1
- G 1×1
- H 1×0
- I 9×1
- J 1×6
- K 1×7
- L 5×1

"2" facts
- A 2×7
- B 4×2
- C 2×5
- D 3×2
- E 8×2
- F 0×2
- G 2×2
- H 5×2
- I 2×3
- J 1×2
- K 2×6
- L 2×9

"3" facts
- A 6×3
- B 3×1
- C 3×6
- D 7×3
- E 3×3
- F 3×0
- G 4×3
- H 3×8
- I 3×2
- J 3×4
- K 5×3
- L 3×9

"4" facts
- A 4×5
- B 1×4
- C 4×8
- D 5×4
- E 4×9
- F 4×0
- G 6×4
- H 3×4
- I 4×2
- J 0×4
- K 4×4
- L 7×4

"5" facts
- A 5×4
- B 5×3
- C 2×5
- D 4×5
- E 5×9
- F 3×5
- G 1×5
- H 5×8
- I 5×2
- J 6×5
- K 5×5
- L 7×5

2. Draw a table like this one. Find the products not given.

×	0	1	2	3	4	5	6	7	8	9
0										
1										
2				8				16		
3										
4			8					28		
5										

3. Find the products.

- A 4×4
- B 4×5
- C 6×2
- D 6×3

- E 8×3
- F 8×4
- G 5×5
- H 5×6

- I 7×2
- J 7×3
- K 9×2
- L 9×4

More practice, page A-18, Set 26

● *What are the facts when 6 and 7 are factors?*

Discussing the Ideas

1. Study the small table.

    A. How can the "2" and the "4" facts help you find the "6" facts?

×	0	1	2	3	4	5	6	7	8	9
2	0	2	4	6	8	10	12	14	16	18
4	0	4	8	12	16	20	24	28	32	36
6									48	

    B. Give the products for A through D in the "6" row of the large table.

2. What other two rows could you use to help you find the products in the "6" row?

×	0	1	2	3	4	5	6	7	8	9
0										
1										
2										
3										
4										
5										
6								A	B	C	D
7								E			
8								F			
9								G			

3. Use the order principle and give the products for E through G in the "6" column.

4. Explain how to use the products given in the table to help find the products for H, I, and J in the "7" row.

×	0	1	2	3	4	5	6	7	8	9
0										
1										
2										
3								21	24	27
4								28	32	36
5										
6										
7								H	I	J
8								K		
9								L		

5. What other two rows could you have used to find the products in the "7" row? Explain.

6. Give the products for K and L.

Using the Ideas

1. Solve the equations.
- A $3 \times 7 = 21 \rightarrow 6 \times 7 = n$
- B $5 \times 8 = 40 \rightarrow 6 \times 8 = n$
- C $2 \times 6 = 12 \rightarrow 4 \times 6 = n$
- D $2 \times 6 = 12 \rightarrow 6 \times 6 = n$

2. Copy and complete the tables.

A
×	5
1	
5	
6	

B
×	6
2	
4	
6	

C
×	7
3	
4	
6	

D
×	8
4	
2	
6	

E
×	9
5	
1	
6	

3. Copy and complete the tables.

A
×	5
3	
4	
7	

B
×	8
3	
4	
7	

C
×	6
5	
2	
7	

D
×	9
2	
5	
7	

E
×	7
6	
1	
7	

4. Find the products.
- A 0×6
- B 1×6
- C 2×6
- D 3×6
- E 4×6
- F 5×6
- G 6×6
- H 7×6
- I 8×6
- J 9×6

5. Find the products.

- A 3×7
- B 6×7
- C 5×7
- D 8×7
- E 9×7
- F 4×7

  7×3 7×6 7×5 7×8 7×9 7×4

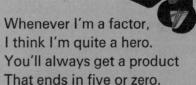

think

Whenever I'm a factor,
I think I'm quite a hero.
You'll always get a product
That ends in five or zero.

WHO AM I?

More practice, page A-19, Set 27

● *What are the facts when 8 and 9 are factors?*

Discussing the Ideas

1. Explain how to use the products in the "3" row and the "5" row to find the products for **A** and **B** in the "8" row.

2. Explain how the fact 4 × 8 = 32 can be used to find 8 × 8.

3. What other two rows could you use to find the products in the "8" row? Explain.

4. A How can the fact 7 × 9 = 63 be used to find 8 × 9?
   B What is the product for **B** in the table?

5. A Why is the product for **C** the same as the product for **B**?
   B What is the product for **C**?

6. A What two rows could you use to find the product for **D** in the "9" row?
   B Find the product for **D** in the table.

7. How can the fact 10 × 9 = 90 be used to find 9 × 9?

8. How can the fact that 3 × 9 = 27 be used to find 9 × 9?

Using the Ideas

1. Copy and complete the tables.

A
×	6
4	
5	
9	

B
×	9
6	
2	
8	

C
×	9
4	
5	
9	

D
×	8
3	
6	
9	

E
×	9
1	
3	
7	

2. Find the products.

A 8 × 0   E 8 × 4   I 8 × 8   M 9 × 2   Q 9 × 6
B 8 × 1   F 8 × 5   J 8 × 9   N 9 × 3   R 9 × 7
C 8 × 2   G 8 × 6   K 9 × 0   O 9 × 4   S 9 × 8
D 8 × 3   H 8 × 7   L 9 × 1   P 9 × 5   T 9 × 9

3. Find the products.

A 4 × 5   I 5 × 3
B 3 × 8   J 7 × 3
C 2 × 9   K 7 × 8
D 0 × 7   L 8 × 8
E 6 × 1   M 6 × 7
F 5 × 6   N 6 × 6
G 4 × 4   O 9 × 7
H 8 × 4   P 9 × 9

think — MULTIPLY BY ?

Multiply me by myself.
You're almost up to fifty.
Though I am a little odd,
I think I'm pretty nifty.

WHO AM I?

4. Find the products.

A 6   B 2   C 8   D 4   E 8   F 6   G 4
  ×2   ×7   ×7   ×5   ×4   ×6   ×4

H 6   I 3   J 5   K 9   L 5   M 5   N 6
  ×7   ×8   ×3   ×7   ×5   ×6   ×3

O 6   P 9   Q 2   R 7   S 9   T 3   U 0
  ×1   ×5   ×9   ×3   ×9   ×4   ×8

More practice, page A-19, Set 28

Short Picture Problems

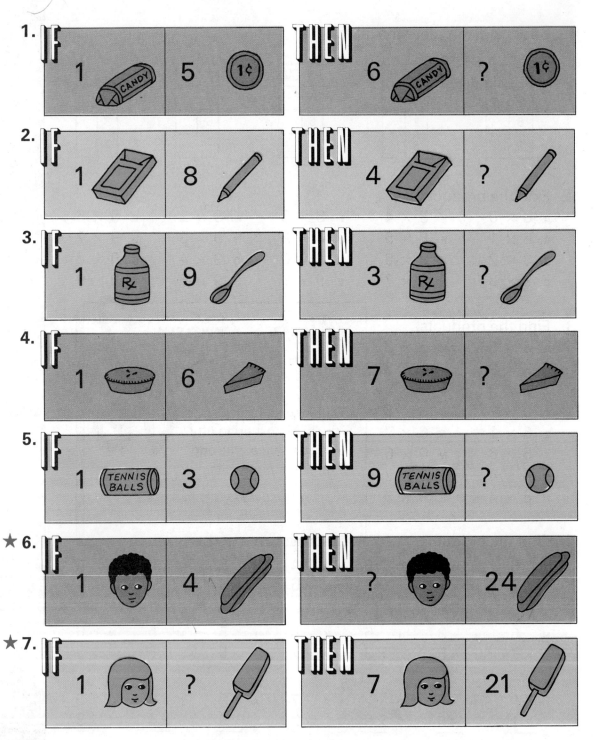

Short Sport Stories

1. 2 basketball teams.
5 players on each team.
How many players?

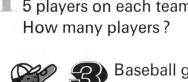

2. 2 basketball teams.
4 cheerleaders for each team.
How many cheerleaders?

3. Baseball game.
6 outs each inning.
9 innings.
How many outs?

6. Tennis.
9 courts.
4 players on each court.
How many players?

4. Red Sox.
3 outs each inning.
9 innings.
How many outs?

5. Basketball game.
4 quarters.
8 minutes each quarter.
How many minutes?

 7. Bowling. 8 balls in each rack. 7 racks. How many bowling balls?

 8. Baseball. 3 strikes, you're out.
8 strikeouts. How many strikes?

9. Football game.
6 points for a touchdown.
5 touchdowns.
How many points?

10. Football game.
6 points for a touchdown.
7 touchdowns.
How many points?

11. Softball. 9 players on each team.
7 teams. How many players?

12. Basketball.
5 fouls, you're out of the game.
4 players out on fouls.
How many fouls for these players?

★13. Football.
6 points for a touchdown.
Bulldogs scored 8 touchdowns and 4 extra points.
What was their score?

157

● *How can patterns help you with the facts?*

Investigating the Ideas

Some "Square Facts"

      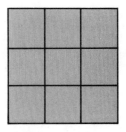

1 × 1 = 1   2 × 2 = 4   3 × 3 = 9

 Can you cut squares from graph paper to show the other square facts up to 9 × 9?

Discussing the Ideas

1. Is (3 × 3) + (4 × 4) = (5 × 5)?

2. Find the number for **n**.
   (6 × 6) + (8 × 8) = (**n** × **n**)

3. The "5" facts are easy. What pattern do you see for the ones digits for the "5" facts?

4. What patterns do you notice in the "9" facts? List all the "9" facts.

```
0 × 5 =  0
1 × 5 =  5
2 × 5 = 10
3 × 5 = 15
4 × 5 = 20
```

```
2 × 9 = 18   (1 + 8 = ?)
3 × 9 = 27   (2 + 7 = ?)
4 × 9 = 36   (3 + 6 = ?)
```

5. Can you list all the whole numbers less than 9 × 9 that are not products for any of the basic facts?

THE FOREST RANGER

Using the Ideas

Last summer Frank spent a week with his Uncle Bill, who is a forest ranger.

1. Frank helped his uncle plant some small pine trees. Frank planted 3 rows of trees. He put 8 trees in each row. How many trees did he plant?

2. Frank learned how to build a fire without matches. Uncle Bill let Frank build the fires for 3 meals each day. Frank stayed 7 days. How many fires did he build?

3. Uncle Bill showed Frank some fish that were to be put in Blue Lake. The fish were in cans. Frank counted 9 cans. There were 8 fish in each can. How many fish were there?

4. Frank helped Uncle Bill put signs on trees. They posted signs on trees along 4 different trails. They put 9 signs on each trail. How many signs did they post?

5. Frank looked at Bald Mountain through a telescope. His uncle said, "It is 4 miles from this tower straight to the top of Bald Mountain. But since you can't fly, it is 5 times as far by the trail." How many miles long is the trail from the tower to the top of the mountain?

More practice, page A-20, Set 29

Practice in Multiplication Facts

1. Give the product for *n* in each equation.
   - A $3 \times 4 = n$
   - B $6 \times 3 = n$
   - C $4 \times 5 = n$
   - D $3 \times 7 = n$
   - E $4 \times 6 = n$
   - F $2 \times 9 = n$
   - G $5 \times 3 = n$
   - H $6 \times 5 = n$
   - I $5 \times 2 = n$
   - J $7 \times 1 = n$
   - K $0 \times 8 = n$
   - L $3 \times 8 = n$

2. How can you use your answers in the equations above to find the missing factors in these equations? Find the missing factors.
   - A $n \times 2 = 10$
   - B $7 \times n = 7$
   - C $6 \times n = 18$
   - D $n \times 5 = 30$
   - E $3 \times n = 12$
   - F $4 \times n = 20$
   - G $n \times 9 = 18$
   - H $5 \times n = 15$
   - I $n \times 7 = 21$
   - J $n \times 8 = 0$
   - K $n \times 8 = 24$
   - L $n \times 6 = 24$

3. Draw a set of 12 dots on your paper. Ring sets of 3 to find how many threes in 12. Write a multiplication equation about this.

4. Draw a set of 20 dots on your paper. Ring sets of 4 to find how many fours in 20. Write a multiplication equation about this.

5. Find the missing factors.
   - A $3 \times n = 9$
   - B $n \times 4 = 16$
   - C $n \times 1 = 7$
   - D $n \times 5 = 15$
   - E $n \times 9 = 0$
   - F $n \times 5 = 25$
   - G $4 \times n = 4$
   - H $n \times 3 = 12$
   - I $2 \times n = 14$
   - J $n \times 2 = 12$
   - K $9 \times n = 18$

think

When you multiply by me,
I really am quite tame.
No matter what the factor is,
The product stays the same.

WHO AM I?

6. Find the missing factors.

	A	B	C	D	E	F	G	H
	▯	3	5	4	4	▯	▯	▯
×	3	▯	▯	▯	▯	5	2	5
	12	9	10	0	12	15	18	25

Solving Story Problems
THE POST OFFICE

1. Jill bought eight 6-cent stamps. How much did they cost?

2. She bought six 8-cent stamps. How much did she spend for them?

3. Jill saw some 1-cent stamps. She counted 3 rows and 9 stamps in each row. How many 1-cent stamps did she see?

4. 6-cent stamps come to the post office in sheets with 10 rows and 10 columns. How many stamps are in each sheet?

5. Jill bought 24 airmail stamps. She bought 4 times as many as Kay. How many airmail stamps did Kay buy?

6. Kay pasted 15 stamps in her collection book. She had 5 rows. How many were there in each row?

7. Which cost more, nine 5-cent stamps or seven 6-cent stamps?

8. Jan spent 32 cents for 8-cent stamps. How many stamps did she buy?

★ 9. Bill needed 29 cents' worth of stamps to mail his package. He had only 5-cent, 6-cent, 7-cent, and 8-cent stamps. He wanted to use only **2 different kinds** of stamps. One way to stamp the package is given in the picture.

   A Find another way, using 7-cent and 8-cent stamps.
   B Find another way, using 5-cent and 8-cent stamps.
   C Find another way.

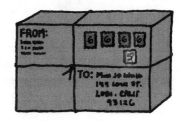

More practice, page A-21, Set 30

● *How are pairing and multiplication related?*

Investigating the Ideas

4 crayons

square   circle

2 types of figures

You are to choose **1 crayon** and **1 figure** to color.
One choice might be green, **circle**. You would do this.

	How many different choices do you think there are?	Show them.

Discussing the Ideas

Sue is getting her mother a birthday present.
She will get her an apron or gloves.
She will put the gift in one of the 3 boxes shown below.

What might Sue's mother find when she unwraps this package?

1. Which gift and box would you choose?

2. Sue finally decided to get the gloves. She put them in the box with the stars. What would your choices have been? Give as many different choices as you can.

3. Solve: | Number of gifts to choose from | Number of boxes to choose from | Number of different choices possible |

$2 \times 3 = n$

Using the Ideas

1. You can have one piece of fruit.

   Apple
   Banana
   Pear
   Orange

   One piece of candy.

   Candycane
   Kiss
   Lollipop

   One of each.

   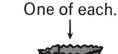

   A How many different pieces of fruit are there?
   B How many different pieces of candy are there?
   C Name all the possible choices that could be in the sack. Use the red letters to stand for each object.
   D Solve: $4 \times 3 = n$

2. 3 flavors of ice cream.

   Vanilla
   Strawberry
   Orange

   3 kinds of syrup.

   One of each.

   A How many different sundaes can you make? List them.
   B Solve: $3 \times 3 = n$

3. Write a multiplication equation for each picture. The small dots help you count the red lines that pair the squares with the circles.

   Example:

   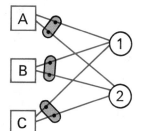

   $2 \times 4 = 8$

   A

   B

   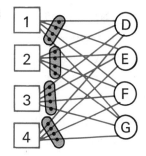

Solving Story Problems

At the CARNIVAL

1. In the Fun House, there are 3 doors (A, B, and C) leading from the Noise Room to the Mirror Room. There are 4 doors (D, E, F, and G) leading from the Mirror Room to the Dark Room.
   - **A** There are 12 ways to get from the Noise Room to the Dark Room. See if you can find them all.
   - **B** Write a multiplication equation about this.

2. There are 2 sizes of ice cream cones. There are 6 flavors to choose from.
   - **A** How many different ice cream cones could you buy?
   - **B** Write a multiplication equation about this.

★ 3. David and Michael each had 1 ticket for rides. There were 4 things they could ride: the merry-go-round, ferris wheel, airplanes, and roller coaster.
   - **A** Give all the ways David and Michael could use their tickets.
   - **B** Write a multiplication equation about this.

Planning a Trip

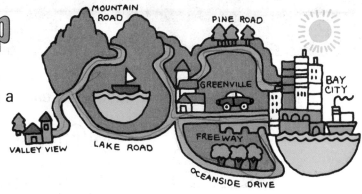

1. Stuart lives in Valley View. When he was helping plan a family trip to Bay City, he thought about these questions. See if you can answer them.

   A In how many ways can we drive from Valley View to Greenville?
   B In how many ways can we drive from Greenville to Bay City?
   C What are the 6 ways to drive from Valley View to Bay City?
   D What multiplication equation can we write about these ideas?

2. Stuart and his sister made plans for the day in Bay City. Here is the list of things they wanted to do.

   A How many afternoon choices are on the list?
   B How many evening choices are on the list?
   C Use the list to give all the possible ways they could spend the day. (You should find 8 ways.)
   D Write a multiplication equation about these ideas.

3. Stuart and his family can take the **helicopter**, the **boat**, or the **bus** from the restaurant to the park gate.
   To get from the gate to the zoo, they can ride the small **train**, ride the **pony**, or **walk**.

   A Stuart wanted to take the boat from the restaurant to the park gate and then take the train from the gate to the zoo. Brenda wanted to take the helicopter and then the pony. Find as many more ways as you can to get from the restaurant to the zoo.
   B Write a multiplication equation about these ideas.

Reviewing the Ideas

1. Write multiplication equations for each exercise.

   A 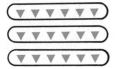   B 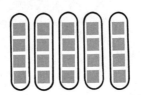   C

   D $5 + 5 + 5 + 5 + 5 + 5$
   E $8 + 8 + 8 + 8$
   F $4 + 4 + 4 + 4 + 4 + 4 + 4$
   G $2 + 2 + 2 + 2 + 2$

   H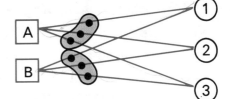

   I

   J

2. Write a multiplication equation about this set by
   A thinking about the rows.
   B thinking about the columns.

   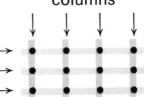

3. Find the missing factors.
   A $6 \times 7 = n \times 6$
   B $9 \times 8 = 8 \times n$
   C $27 \times 56 = n \times 27$
   D $98 \times 76 = 76 \times n$

4. Complete each statement.
   A The product of any number and one is __?__.
   B The product of any number and zero is __?__.

5. Solve the equations.
   A $5 \times 4 = (3 \times 4) + (n \times 4)$
   B $8 \times 3 = (4 \times 3) + (n \times 3)$
   C $7 \times 5 = (6 \times 5) + (n \times 5)$

5. Find the products.

   A. 6 × 0
   B. 1 × 8
   C. 5 × 3
   D. 2 × 6
   E. 0 × 9
   F. 5 × 5
   G. 7 × 1
   H. 4 × 3
   I. 2 × 7
   J. 5 × 4
   K. 2 × 5
   L. 5 × 6
   M. 2 × 9
   N. 5 × 7
   O. 2 × 8
   P. 6 × 2
   Q. Since 9 × 6 = 54, we know 6 × 9 = *n*.
   R. Since 6 × 7 = 42, we know 7 × 6 = *n*.
   S. Since 8 × 6 = 48, we know 6 × 8 = *n*.
   T. Since 9 × 8 = 72, we know 8 × 9 = *n*.
   U. Since 5 × 5 = 25, we know 6 × 5 = *n*.
   V. Since 6 × 6 = 36, we know 7 × 6 = *n*.
   W. Since 7 × 7 = 49, we know 8 × 7 = *n*.
   X. Since 8 × 8 = 64, we know 9 × 8 = *n*.

6. Find the products.

   A. 3 × 3
   B. 7 × 8
   C. 3 × 4
   D. 6 × 7
   E. 4 × 5
   F. 7 × 7
   G. 3 × 9
   H. 8 × 8
   I. 4 × 7
   J. 6 × 8
   K. 5 × 8
   L. 4 × 4
   M. 3 × 8
   N. 4 × 9
   O. 3 × 6
   P. 6 × 6
   Q. 4 × 6
   R. 7 × 9
   S. 3 × 5
   T. 9 × 9
   U. 6 × 9
   V. 3 × 7
   W. 4 × 8
   X. 8 × 9

7. Complete the sentence. Then find the product.

   A. For 3 sets of 10, we write ||||. 3 × 10 = *n*
   B. For 5 sets of 10, we write ||||. 5 × 10 = *n*
   C. For 8 sets of 10, we write ||||. 8 × 10 = *n*
   D. For 7 sets of 10, we write ||||. 7 × 10 = *n*

think

If it's products you must find,
You'll think I'm quite a guy.
I don't cause a single change
When I just multiply.

WHO AM I?

8. Solve the equations.

   A. (2 × 7) + (3 × 7) = *n*
   B. (8 × 7) + (2 × 7) = *n*
   C. (5 × 9) + (5 × 9) = *n*
   D. (6 × 8) + (4 × 8) = *n*
   E. (7 × 7) + (3 × 7) = *n*
   F. (9 × 8) + (1 × 8) = *n*

Keeping in Touch with Addition / Subtraction / Place value

1. Find the sums.
   - A. 3 + 9
   - B. 8 + 3
   - C. 5 + 7
   - D. 7 + 7
   - E. 8 + 5
   - F. 4 + 9
   - G. 4 + 7
   - H. 9 + 6
   - I. 7 + 8
   - J. 8 + 9
   - K. 6 + 8
   - L. 9 + 5
   - M. 7 + 9
   - N. 9 + 9
   - O. 8 + 8
   - P. 7 + 6

2. Find the differences.
   - A. 11 − 9
   - B. 12 − 8
   - C. 16 − 7
   - D. 11 − 8
   - E. 12 − 3
   - F. 13 − 8
   - G. 14 − 9
   - H. 14 − 7
   - I. 11 − 6
   - J. 17 − 8
   - K. 13 − 7
   - L. 12 − 5
   - M. 15 − 7
   - N. 16 − 8
   - O. 18 − 9
   - P. 15 − 9

3. Find the sums.
   - A. 62 + 23
   - B. 35 + 24
   - C. 16 + 51
   - D. 328 + 160
   - E. 457 + 132
   - F. 615 + 204

4. Find the differences.
   - A. 89 − 23
   - B. 68 − 51
   - C. 586 − 212

5. Find the sums.
   - A. 38 + 27
   - B. 65 + 25
   - C. 73 + 18
   - D. 75 + 86
   - E. 38 + 95
   - F. 67 + 67

think

The product is even
When I multiply.
On words like bicycle,
My nickname is "bi."

WHO AM I?

6. Find the differences.

   A 36   B 43   C 52   D 58   E 64   F 80
     −17   −24   −17   −29   −27   −68

   G 123   H 144   I 163   J 150   K 131   L 155
     −47   −58   −76   −68   −75   −67

7. Find the sums.
   A 60 + 9   E 500 + 40
   B 200 + 40 + 7   F 40 + 200 + 3
   C 600 + 10 + 4   G 90 + 2 + 800
   D 900 + 2   H 9 + 900 + 90

think

If you add me to myself
Then multiply by two,
It's 12 plus 4 you're
  sure to get.
You need no other clue.

WHO AM I?

8. Find the sums.
   A 3000 + 600 + 20 + 8
   B 4000 + 200 + 80 + 9
   C 200 + 6 + 4000 + 30
   D 50 + 6000 + 200 + 6

9. Study the example. Then write each number as shown in the example.
   Example: 8295 = 8000 + 200 + 90 + 5
   A 654   B 892   C 4347   D 8126   E 87,265   F 27,615

★ 10. Copy each problem. Give the missing digit for each ▓.

   A ▓▓   B ▓3   C 84   D ▓4▓   E 63▓
     −27   −36   −3▓   −237   −207
     ───   ───   ───   ────   ────
      12   5▓   ▓7   6▓2   ▓▓8

You are invited to explore

ACTIVITY CARD 7
Page 312

7 Division

● *What is division?*

Investigating the Ideas

24 counters — Put 4 in each cup. — How many cups?

Number of counters	Counters in each cup	Number of cups	Division equation
24	4	?	24 ÷ 4 = ?
24	2	?	24 ÷ 2 = ?
24	3	?	24 ÷ 3 = ?
24	?	?	

 Can you use counters and containers to help you complete the chart?

Discussing the Ideas

1. Give the missing numbers. Write a division equation for B.

   We see   We think   We write

   A 12 tomatoes — 4 in each box — There are ||||| fours in 12. — 12 ÷ 4 = 3
   (Read: 12 divided by 4 equals 3.)

   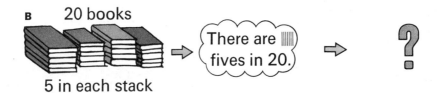

   B 20 books — 5 in each stack — There are ||||| fives in 20. — ?

2. Make up an example like those above.

Using the Ideas

1.
   A How many keys in all?
   B How many on each ring?
   C How many rings?
   D How many threes in 12?
   E Solve: $12 \div 3 = n$

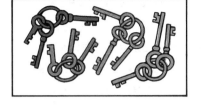

2.
   A How many letters in all?
   B How many in each set?
   C How many sets?
   D How many threes in 15?
   E Solve: $15 \div 3 = n$

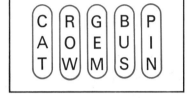

3.
   A How many shoes in all?
   B How many in each pair?
   C How many pairs?
   D How many twos in 10?
   E Solve: $10 \div 2 = n$

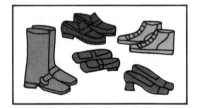

4.
   A How long is the green strip?
   B How many of the red strips are needed to match the green strip?
   C How many twos in 6?
   D Solve: $6 \div 2 = n$

When you solved the equations above, you found the **quotient**.

5. Draw 21 dots. Ring as many sets of 3 as you can.
   A How many sets of 3 did you find?
   B How many threes in 21?
   C Find the quotient: $21 \div 3 = n$

6.
   A Draw 24 dots. Ring as many sets of 4 as you can.
   B Solve: $24 \div 4 = n$

● Can "rectangular sets" help you find quotients?

Investigating the Ideas

How long do you think a rectangle with 28 squares would be if it were 4 units wide?

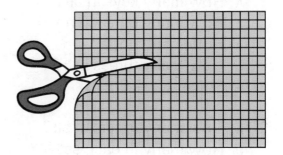

 Can you cut from graph paper a rectangle with 30 squares that is 5 units wide?

Discussing the Ideas

1. What division equation can you write about the rectangle you cut out?

2. A Can you find some other rectangles that have 30 squares?
   B Write division equations for each one you find.

3. Give the missing numbers and solve the equations.

	Rectangle	Width	Length	Division equation
A	12 squares	2 units		$12 \div 2 = n$
B	15 squares	3 units		$15 \div 3 = n$
C	18 squares		6 units	$18 \div 6 = n$
D	16 squares	4 units		$16 \div 4 = n$

4. Explain how you can use this "rectangular set" to get 2 division equations.

$3 \times 8 = 24$

Using the Ideas

1. Look at the set. Then answer the question.

   A
   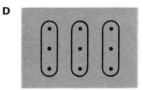
   How many sets of 4 in a set of 8?

   B

   How many sets of 3 in a set of 12?

   C
   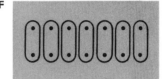
   How many sets of 5 in a set of 15?

   D
   How many threes are in 9?

   E
   How many sixes are in 24?

   F
   How many twos are in 14?

2. Solve the equations. Your work in exercise 1 will help you.

   A $8 \div 4 = n$ C $15 \div 5 = n$ E $24 \div 6 = n$
   B $12 \div 3 = n$ D $9 \div 3 = n$ F $14 \div 2 = n$

Short Stories

1. 15 chocolates. 3 in each row. How many rows?

2. 20 boys. 5 on a team. How many teams?

3. 18 letters. 3 in each word. How many words?

4. 24 books. 6 in each stack. How many stacks?

5. 15 marbles. 5 in a sack. How many sacks?

6. 16 white rats. 4 per cage. How many cages?

7. 30 bottles of pop. 6 bottles in a carton. How many cartons?

8. 25 pounds of sugar. Put in 5-pound bags. How many bags?

173

● Can you use subtraction to help you find quotients?

Discussing the Ideas

1. Think about removing the dots in the ring.
   Then solve the subtraction equations.

   A
   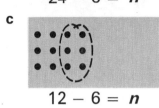
   $24 - 6 = n$

   B
   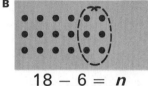
   $18 - 6 = n$

   C
   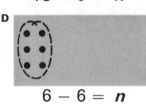
   $12 - 6 = n$

   D
   $6 - 6 = n$

2. A Study exercise 1 and tell how many sixes are in 24.

   B Solve this equation. $24 \div 6 = n$

3. Study each example. Solve and explain the equation.

   A
   We can use addition to help us find products. 4×5
   Since $5 + 5 + 5 + 5 = 20$, we know that $4 \times 5 = n$.

   B
   We can use subtraction to help us find quotients. $18 \div 6$

   $$\begin{array}{r} 18 \\ -6 \\ \hline 12 \end{array} \quad \begin{array}{r} 12 \\ -6 \\ \hline 6 \end{array} \quad \begin{array}{r} 6 \\ -6 \\ \hline 0 \end{array}$$

   Since we subtracted 6 three times, we know that $18 \div 6 = n$.

4. Explain how you could find $36 \div 3$ by subtracting.
   Solve: $36 \div 3 = n$

Using the Ideas

1. **A** Find these differences.

24	20	16	12	8	4
−4	−4	−4	−4	−4	−4

   B How many times did you subtract 4?
   C How many fours are in 24?
   D Write a division equation about this.

2. **A** Find these differences.

21	18	15	12	9	6	3
−3	−3	−3	−3	−3	−3	−3

   B How many times did you subtract 3?
   C How many threes are in 21?
   D Write a division equation about this.

3. **A** Find these differences.

30	25	20	15	10	5
−5	−5	−5	−5	−5	−5

   B How many times did you subtract 5?
   C How many fives are in 30?
   D Write a division equation about this.

4. Solve the equations.

   A $15 - 5 = n$ **B** $28 - 7 = n$
       $10 - 5 = n$ $21 - 7 = n$
        $5 - 5 = n$ $14 - 7 = n$
       $15 \div 5 = n$ $7 - 7 = n$
                       $28 \div 7 = n$

★ 5. Find the quotients.
   A $28 \div 2$ **C** $48 \div 4$
   B $42 \div 3$ **D** $75 \div 5$

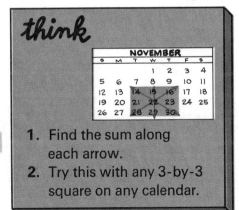

think

1. Find the sum along each arrow.
2. Try this with any 3-by-3 square on any calendar.

● Can the number line help you think about division?

Investigating the Ideas

Use your centimeter ruler as a number line. Starting at 24, it takes 8 jumps of three to get to zero.
There are 8 threes in 24.

$$24 \div 3 = 8$$

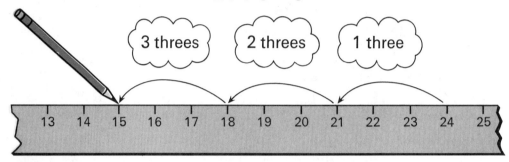

 How many different division equations can you write to show other jumps you can make to get from 24 to zero?

Discussing the Ideas

1. What division fact will you discover if you start at 18 on your ruler and jump back 2 at a time?

2. Explain why starting at 16 and jumping by threes will not give you a division fact.

3. Explain how each figure below helps you solve $6 \div 2 = n$.

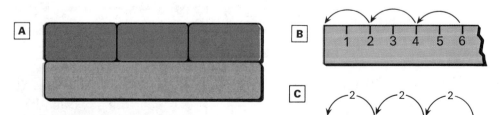

Using the Ideas

1.

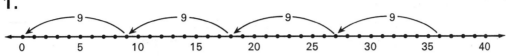

A How many nines in 36?   B $36 \div 9 = n$

2.

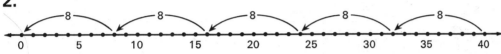

A How many eights in 40?   B $40 \div 8 = n$

3.

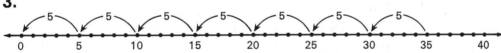

A How many fives in 35?   B $35 \div 5 = n$

4. Draw a number line to help you find each of these.

A $14 \div 2$ B $12 \div 4$ C $20 \div 5$ D $16 \div 2$

Solving Story Problems

1. Mr. Smith paid $15 for the children's baseball tickets. They were $3 each. How many children went to the ball game with Mr. Smith?

2. The Blue Sox scored 2 runs in each inning until the scoreboard read How many innings had they played then?

BLUE SOX	8
GREEN SOX	0

★ 3. Ted lived 15 miles from the ball park. Mr. Smith drove at the rate of 3 miles each 5 minutes. How long did it take to get from the ball park to Ted's house?

More practice, page A-22, Set 31

● Can you find quotients by finding missing factors?

Investigating the Ideas

Cut from graph paper as many different rectangles that have 36 squares as you can.

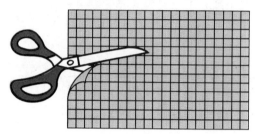

Number of squares in a row		Number of rows		Number of squares in the rectangle
6	×	?	=	36
9	×	?	=	36
?	×	3	=	36
18	×	?	=	36
?	×	1	=	36

 Can you complete this table with the help of your rectangles?

Discussing the Ideas

1. On Friday, the children checked their arithmetic papers. Peter checked Judy's paper. Under one exercise, he wrote ⟶

   P   F   F
   24 ÷ 4 = 5 ✗
   When 5 and 4 are factors the product is 20, not 24.

   A Explain what Peter was trying to tell Judy.
   B What do you think Peter would say about this equation?
   $18 \div 3 = 7$

2. Find the missing factors and quotients. Explain your answers.

   A
   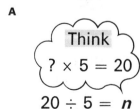
   Think
   ? × 5 = 20

   $20 \div 5 = n$

   B

   Think
   ? × 6 = 6

   $6 \div 6 = n$

   C
   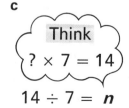
   Think
   ? × 7 = 14

   $14 \div 7 = n$

Using the Ideas

1. Find the missing factors.

   A *To find this quotient, I think ? × 5 = 15.*   15 ÷ 5

   B *To find this quotient, I think ? × 3 = 12.*   12 ÷ 3

2. Find the quotients.

   A Think ? × 2 = 6   6 ÷ 2 = *n*

   B Think ? × 3 = 9   9 ÷ 3 = *n*

   C Think ? × 4 = 8   8 ÷ 4 = *n*

3. Find the missing factors and quotients.
   A Since *n* × 4 = 8, we know that 8 ÷ 4 = *n*.
   B Since *n* × 3 = 9, we know that 9 ÷ 3 = *n*.
   C Since *n* × 5 = 15, we know that 15 ÷ 5 = *n*.
   D Since *n* × 3 = 15, we know that 15 ÷ 3 = *n*.
   E Since *n* × 5 = 20, we know that 20 ÷ 5 = *n*.

4. Find the quotients. Use multiplication to check your answers.
   A 24 ÷ 8 = *n*
   B 30 ÷ 6 = *n*
   C 27 ÷ 9 = *n*
   D 12 ÷ 3 = *n*
   E 27 ÷ 3 = *n*
   F 21 ÷ 3 = *n*
   G 16 ÷ 8 = *n*

think 18 2

Our names you're sure to find
If you use this little clue.
Our product is 18,
While our quotient's only 2.

WHO ARE WE?

More practice, page A-23, Set 32

● *Do you understand division?*

Discussing the Ideas

1. For each example in the chart, solve the equation. Explain your answer.

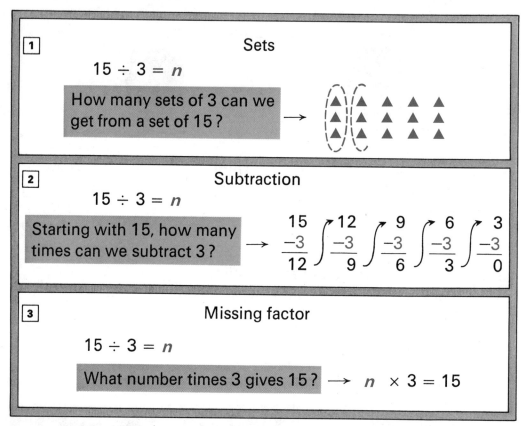

2. Explain how you could use your 2-strip and 10-strip to find $10 \div 2$.

3. How could you use a centimeter ruler to find $27 \div 3$?

4. If you know $7 \times 8 = 56$, what two division facts can you give?

Using the Ideas

1. Write a division equation for each exercise.

   A 　　B 　　C

2. Write a division equation for each exercise.

   A $\begin{array}{r}27\\-9\\\hline 18\end{array}$　$\begin{array}{r}18\\-9\\\hline 9\end{array}$　$\begin{array}{r}9\\-9\\\hline 0\end{array}$　　B $\begin{array}{r}35\\-7\\\hline 28\end{array}$　$\begin{array}{r}28\\-7\\\hline 21\end{array}$　$\begin{array}{r}21\\-7\\\hline 14\end{array}$　$\begin{array}{r}14\\-7\\\hline 7\end{array}$　$\begin{array}{r}7\\-7\\\hline 0\end{array}$

3. Solve the equations.　　4. Now find the quotients.

   A $n \times 6 = 12$　⟶　A $12 \div 6 = n$
   B $n \times 3 = 9$　⟶　B $9 \div 3 = n$
   C $n \times 4 = 8$　⟶　C $8 \div 4 = n$
   D $n \times 2 = 10$　⟶　D $10 \div 2 = n$
   E $n \times 3 = 12$　⟶　E $12 \div 3 = n$
   F $n \times 5 = 25$　⟶　F $25 \div 5 = n$

5. In each exercise, tell how many bags of marbles.

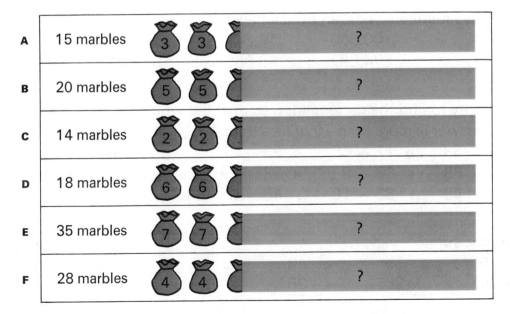

Keeping in Touch with

Addition
Subtraction
Inequalities

1. Write 2 addition and 2 subtraction equations for each figure.

   A 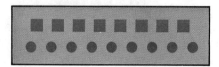   B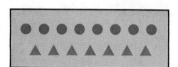

2. Give the sums as quickly as you can.

   A 5 + 4   E 5 + 5   I 8 + 9   M 5 + 6   Q 7 + 6
   B 6 + 2   F 8 + 3   J 3 + 7   N 7 + 5   R 9 + 9
   C 9 + 1   G 7 + 7   K 8 + 2   O 8 + 8   S 6 + 8
   D 4 + 6   H 6 + 7   L 3 + 5   P 4 + 9   T 7 + 8

3. Give the differences as quickly as you can.

   A 8 − 3   E 13 − 8   I 15 − 9   M 14 − 6   Q 12 − 7
   B 9 − 2   F 15 − 7   J 13 − 5   N 9 − 4   R 15 − 8
   C 6 − 5   G 17 − 9   K 14 − 5   O 16 − 7   S 10 − 3
   D 11 − 4   H 16 − 8   L 10 − 6   P 18 − 9   T 13 − 7

4. List these numbers in order, from smallest to largest.

   68,423   100,000   16,842   50,000   68,433
   49,375   674,897   49,365   684,230   68,300

5. Do not try to find the correct answer. Just tell whether each answer is more than 70 or less than 70.

   A 37 + 38   D 17 + 58   G 93 − 18   J 95 − 26   M 50 + 21
   B 27 + 38   E 83 − 8   H 93 − 28   K 93 − 19   N 80 − 11
   C 27 + 48   F 83 − 18   I 46 + 27   L 15 + 54   O 90 − 19

6. Find the sums.

   A   35   B   35   C   48   D   57   E   68   F   73
      +24   +25   +26   +19   +24   +14

7. Find the sums.

A 66 +27	B 99 +4	C 88 +7	D 76 +66	E 35 +48	F 87 +32
G 93 +27	H 69 +25	I 46 +58	J 57 +48	K 76 +64	L 19 +27

8. Find the differences.

A 83 −22	B 92 −33	C 57 −34
D 61 −59	E 122 −34	F 74 −68
G 182 −93	H 65 −37	I 136 −48

think

Dave is three times as old as Sue. In three years, he will be twice as old as Sue. How old are Dave and Sue now?

9. Find the sums and differences.

A 35 +28	B 72 −43	C 68 +78	D 39 −19	E 92 −15
F 67 +67	G 95 +78	H 60 −17	I 34 −19	J 78 +62

★ 10. Find the sums and differences.

A 345 +167	B 6291 +1963	C 642 −285	D 1354 −678	E 1876 +695	F 3604 −538

You are invited to explore

ACTIVITY CARD 8
Page 313

Can multiplication facts help you with division facts?

First find the products in exercise 1. You can then use these facts to help you do exercise 2.

1. Find the products.

   A $2 \times 7 = n$ F $4 \times 4 = n$ K $6 \times 6 = n$ P $5 \times 3 = n$
   B $9 \times 2 = n$ G $7 \times 5 = n$ L $3 \times 5 = n$ Q $4 \times 7 = n$
   C $5 \times 4 = n$ H $2 \times 6 = n$ M $1 \times 9 = n$ R $6 \times 0 = n$
   D $6 \times 3 = n$ I $5 \times 5 = n$ N $4 \times 6 = n$ S $5 \times 8 = n$
   E $8 \times 4 = n$ J $9 \times 3 = n$ O $3 \times 8 = n$ T $6 \times 5 = n$

2. Find the missing factors. Use exercise 1 to check your answers.

   A $n \times 8 = 40$ I $n \times 7 = 28$ Q $n \times 3 = 15$ S $6 \times n = 18$
   B $n \times 3 = 27$ J $6 \times n = 36$ R $n \times 5 = 30$ T $n \times 5 = 15$
   C $4 \times n = 24$ K $8 \times n = 32$
   D $6 \times n = 0$ L $n \times 9 = 9$
   E $n \times 2 = 18$ M $n \times 5 = 25$
   F $n \times 8 = 24$ N $5 \times n = 20$
   G $4 \times n = 16$ O $n \times 5 = 35$
   H $2 \times n = 14$ P $2 \times n = 12$

3. Find the quotients.

   A $14 \div 2 = n$ K $18 \div 2 = n$
   B $35 \div 5 = n$ L $36 \div 6 = n$
   C $18 \div 6 = n$ M $40 \div 8 = n$
   D $12 \div 2 = n$ N $30 \div 5 = n$
   E $27 \div 3 = n$ O $28 \div 7 = n$
   F $16 \div 4 = n$ P $0 \div 6 = n$
   G $32 \div 8 = n$ Q $25 \div 5 = n$
   H $15 \div 5 = n$ R $24 \div 8 = n$
   I $24 \div 4 = n$ S $15 \div 3 = n$
   J $9 \div 9 = n$ T $20 \div 5 = n$

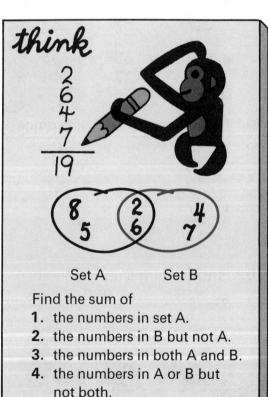

Set A   Set B

Find the sum of
1. the numbers in set A.
2. the numbers in B but not A.
3. the numbers in both A and B.
4. the numbers in A or B but not both.

Function Machine Problems

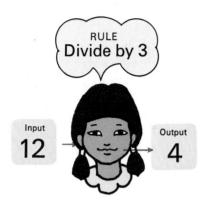

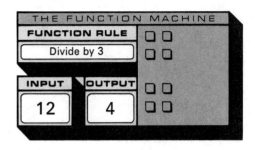

Think about the function machine and give the missing numbers.

1. Function Rule: Divide by 2

	Input	Output
	8	4
	6	3
A	10	
B	4	
C	12	

2. Function Rule: Divide by 3

	Input	Output
	6	2
A	15	
	12	4
B	9	
C	3	

3. Function Rule: Divide by 4

	Input	Output
	12	3
A	8	
B	16	
C	4	
	20	5

4. Function Rule:
A: Divide by ▊

	Input	Output
	5	1
B	10	
	15	3
	20	4
C	25	

5. Function Rule: Divide by 10

	Input	Output
	20	2
A	60	
	50	5
B	80	
C		4

6. Function Rule: Divide by 2

	Input	Output
	12	6
A		5
B		7
C		4
	16	8

More practice, page A-23, Set 33

● How are division and sets related?

Investigating the Ideas

Get some counters, paper cups, and a collection box.

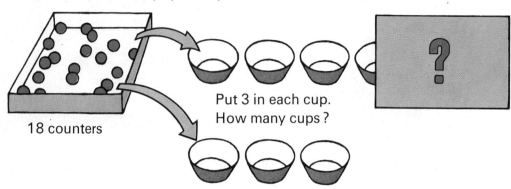

18 counters

Put 3 in each cup.
How many cups?

Divide equally in 3 cups.
How many in each cup?

 Can you use your counters to help you complete the chart?

Total number	Number in each cup	Number of cups	Division equation
18	3	?	18 ÷ 3 = ?
18	?	3	18 ÷ 3 = ?
18	?	6	||||||||||||||||||||||
18	?	2	||||||||||||||||||||||

Discussing the Ideas

1. You can use division to find **how many sets.**
   Solve the equation to find out.

    ➡ How many stacks of 4 can you make? ➡ 12 ÷ 4 = *n*

   12 books

2. You can use division to find **how many in each set.**
   Solve the equation to find out.

    ➡ If the cookies are divided equally among 3 children, how many does each child get? ➡ 15 ÷ 3 = *n*

   15 cookies

Using the Ideas

1. Write a division equation for each exercise.

   A 6 apples

   B 10 cents

   C 8 flowers

   A How many does each child get if you divide the apples equally among 3 children?

   B How many children can have 2 cents each?

   C How many are in each vase if they are divided equally in 2 vases?

2. In each part, the marbles are equally divided in the bags.

A	18 marbles	3, 3, ?	How many bags?
B	16 marbles	?, ?, ?, ?	How many in each bag?
C	24 marbles	4, 4, ?	How many bags?
D	30 marbles	?, ?, ?, ?, ?, ?	How many in each bag?
E	28 marbles	4, 4, ?	How many bags?
F	27 marbles	9, 9, ?	How many bags?
G	40 marbles	?, ?, ?, ?, ?	How many in each bag?

More practice, page A-24, Set 34

● How well do you understand division?

Investigating the Ideas

Suppose there are 30 chairs in a room.

1. How many rows of could you make?
2. How many rows of could you make?
3. How many rows of could you make?
4. How many chairs in each row if you divide them equally in ten rows?

| ? | Can you use a set of counters and show your answer to each question? | Draw pictures to show how you arranged your counters. |

Discussing the Ideas

1. What division fact can you give for each of the questions above?

2. Answer the questions about the set of children.
   A How many groups of 9 can be found?
   B How many at each table if they are divided equally among 6 tables?
   C How many girls are there if there are as many boys as girls?
   D How many teams of 3 can be formed for a spelling contest?

18 children

3. How long is each jump?

4. How many jumps of 3 will it take to get to 0?

Using the Ideas

Solve the picture problems.

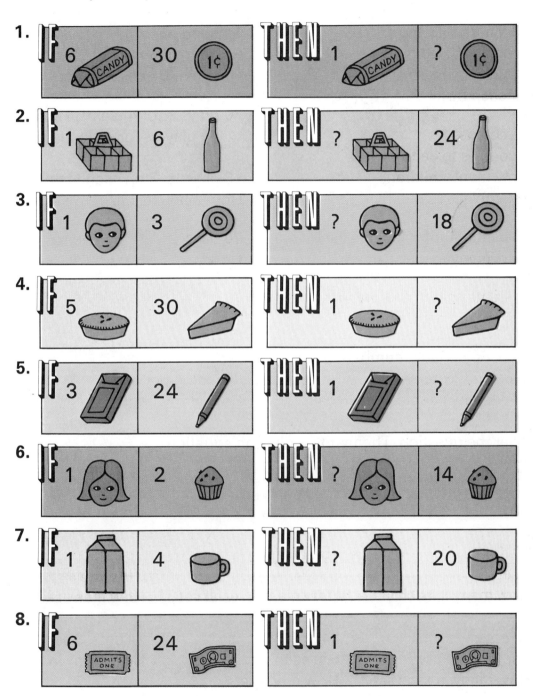

Solving Story Problems

1. Answer the question. Then write a division equation about the story.

   A. 20 marbles. Same number to each of 5 boys. How many marbles for each boy?

   B. 15 cookies. 3 cookies to each child. How many children get cookies?

   C. 30 cents. Apples 6 cents each. How many apples can we buy?

   D. 24 cents spent for 3 oranges. How much per orange?

   E. 24 people. Same number in each of 6 cars. How many people in each car?

   F. 14 children. Same number of girls as boys. How many of each?

   G. 20 players for 4 teams. How many players for each team?

   H. 18 players. 3 on each team. How many teams?

   I. 28 days. 7 days per week. How many weeks?

   J. 16 minutes. 2 minutes per mile. How many miles?

   K. 36 feet. 4 per dog. How many dogs?

2. Answer the question. Then write a division equation.

   A. 12 dots. 4 in each set. How many sets?

   B. 12 dots. 2 sets. How many in each set?

   C. 15 dots. 3 in each set. How many sets?

   D. 10 dots. 5 sets. How many in each set?

   E. 15 dots. 3 sets. How many in each set?

   F. 18 dots. 3 in each set. How many sets?

   G. 16 dots. 4 sets. How many in each set?

   H. 24 dots. 6 in each set. How many sets?

   I. 24 dots. 8 sets. How many in each set?

At the Dairy

1. Cartons of milk are put in boxes, with 6 cartons in each row. If a box holds 24 cartons of milk, how many rows are there?

2. A machine fills 28 half-gallon milk cartons every 4 minutes. How many cartons does it fill in one minute?

3. Cartons of ice cream are placed in large wire racks before they are put in the freezer. Each rack holds 48 cartons in 6 rows. How many cartons are in each row?

4. When Susan's class went to the dairy, they went into the pasteurizing room in groups of 6. There are 30 children in Susan's class. How many groups of 6 did they have?

5. On the way back to school, 5 children rode in each car. How many cars did they need?

6. Miss Smith, the teacher, asked the children to write a story about their trip to the dairy. Don decided to make up some problems for his story. See if you can work them.

How many cartons are in each row?

How many cartons are in each row?

More practice, page A-26, Set 36

Reviewing the Ideas

1. Find the quotients.

   A Since $4 \times 9 = 36$, we know that $36 \div 9 = n$.
   $36 \div 4 = n$.

   B Since $6 \times 8 = 48$, we know that $48 \div 8 = n$.
   $48 \div 6 = n$.

   C Since $9 \times 7 = 63$, we know that $63 \div 9 = n$.
   $63 \div 7 = n$.

   D Since $5 \times 8 = 40$, we know that $40 \div 5 = n$.
   $40 \div 8 = n$.

2. Find the missing factors.
   - A $n \times 3 = 12$
   - C $n \times 6 = 24$
   - E $n \times 3 = 15$
   - G $n \times 3 = 27$
   - B $7 \times n = 14$
   - D $3 \times n = 24$
   - F $8 \times n = 32$
   - H $6 \times n = 36$

3. Find the quotients.
   - A $27 \div 3 = n$
   - C $14 \div 7 = n$
   - E $12 \div 3 = n$
   - G $32 \div 8 = n$
   - B $24 \div 3 = n$
   - D $24 \div 6 = n$
   - F $36 \div 6 = n$
   - H $15 \div 3 = n$

4. $24 \rightarrow 21 \rightarrow 18 \rightarrow 15 \rightarrow 12 \rightarrow 9 \rightarrow 6 \rightarrow 3$
   $-3 \quad -3 \quad -3 \quad -3 \quad -3 \quad -3 \quad -3 \quad -3$
   $21 \quad 18 \quad 15 \quad 12 \quad 9 \quad 6 \quad 3 \quad 0$

   A How many times was 3 subtracted? B How many threes are in 24?
   C Write a division equation about this.

5. Write a division equation for each number-line picture.

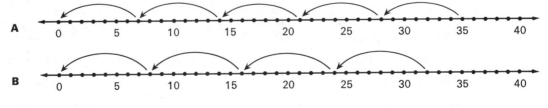

Short Stories

1 18 dots.
3 in each column.
How many columns?

2 15 children.
5 in each car.
How many cars?

3 48 books. 8 shelves
(same number on each).
How many books on each shelf?

4 Rode bicycle 32 miles.
Traveled 4 miles each hour.
How many hours?

6 50 marbles. 5 sacks
(same number in each).
How many marbles
in each sack?

5 14 pieces of candy in 2 hands.
Same number in each hand.
How many pieces in each hand?

7 27 sails.
3 sails on each boat.
How many boats?

8 36 players. 9 on each team. How many teams?

9 Have 35 cents.
Apples 7 cents each.
How many apples can we buy?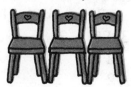

10 36 pieces of pie.
6 pieces in each pie.
How many pies?

11 6 rows of chairs. 42 chairs.
How many chairs in each row?

12 28 jet plane engines.
4 engines on each plane.
How many jet planes?

Keeping in Touch with — Addition, Subtraction, Multiplication, Division, Place value, Inequalities

1. For each pair of numbers, write the larger one on your paper.

   A 84   B 807   C 8263   D 4003   E 8327
      64   811   7379   4010   8309
   F 43,005   G 40,040   H 60,003   I 648,356   J 260,009
      43,004   40,050   60,012   647,356   260,101

2. Find the sums and differences.

   A 38   B 72   C 65   D 39   E 56   F 120   G 148
     +46   −39   +43   −19   +9   −7   +32

   H 74   I 93   J 65   K 127   L 88   M 702   N 803
     −56   +87   −28   −68   +88   −24   −56

3. Answer **T** (true) or **F** (false).
   A 7 + 6 is one more than 6 + 6.
   B 7 × 6 is six more than 6 × 6.
   C 26 + 27 > 50
   D Ten hundreds are one thousand.
   E One hundred thousand is one million.
   F 80 ÷ 8 = 10
   G 90,000 × 9 = 10,000
   H One thousand is one hundred tens.
   I 56 − 28 < 30
   J 9 × 5 is nine more than 8 × 5.
   ★ K 12 × 14 is one less than 13 × 13.
   ★ L 15 × 15 is one more than 14 × 16.
   ★ M 37 + 35 > 70 and 48 + 54 < 100
   ★ N 3486 + 3486 = 3485 + 3487
   ★ O 7 + 5 < 12 or 7 + 5 = 12 or 7 + 5 > 12

think

A missing factor
I cannot be
Whenever the product
Is larger than me.

WHO AM I?

4. Tell how many beads in each exercise. Pretend that the large red cans hold 1000 beads each, the gray cans hold 100 beads each, and the small red cans hold 10 beads each.

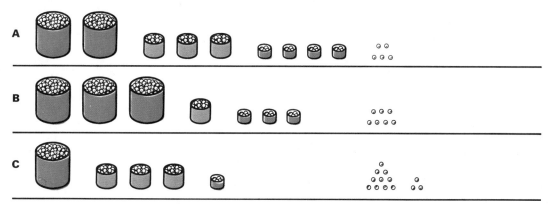

5. Find the products.

A 2 × 6	D 7 × 2	G 6 × 5	J 8 × 4	M 5 × 9	P 4 × 7
B 3 × 4	E 9 × 3	H 2 × 8	K 3 × 8	N 9 × 2	Q 3 × 6
C 4 × 9	F 6 × 6	I 5 × 5	L 7 × 3	O 5 × 7	R 8 × 5

6. No numbers are given in these exercises. You are to tell whether you would add, subtract, multiply, or divide to find each answer.

  A John has ||||| marbles and Bill has ||||| marbles. How many marbles do they have together?

  B In exercise A, how many more marbles does John have than Bill?

  C Mike arranged ||||| chairs in ||||| rows. How many chairs in each row?

  D Jim has ||||| pages of stamps with ||||| stamps on each page. How many stamps does he have on these pages?

  E There are ||||| pieces of candy to be passed among ||||| children. All the children get the same number of pieces. How many pieces does each child get?

You are invited to explore   ACTIVITY CARD 9   Page 313

8 Geometry

● *Can you draw parallel lines?*

Investigating the Ideas

Fold a strip of paper in half.

Fold it in half again.

Fold it in half a third time.

 How many fold lines will you have on your strip?

Discussing the Ideas

1. The fold lines on your strip suggest **parallel lines**. Can you tell what you think parallel lines are?

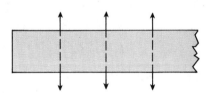

2. These pictures suggest parallel lines. Can you think of other things that suggest parallel lines?

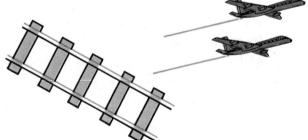

3. **Parallel segments** lie on parallel lines. Can you find some objects in your classroom that suggest parallel segments?

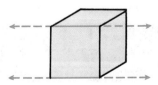

Using the Ideas

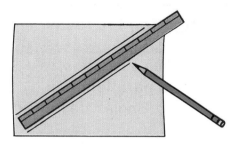

1. You can draw a pair of parallel lines by drawing on each side of your ruler.
   - A Draw two parallel lines using this method.
   - B With your ruler draw two more parallel lines that cross (intersect) the first two lines.

2. Make folds like these and draw along the parallel lines that are formed. Can you fold paper another way to make parallel lines?

   Crease carefully.

   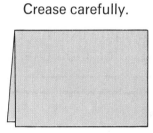

   These edges must come together.

   Fold here.

   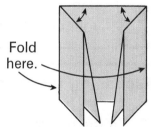

3. Name the pair of lines in the figure that seem to be parallel.

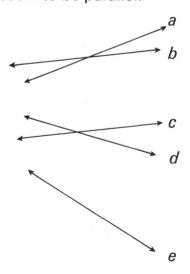

think

Suppose you draw a large right triangle with legs the same length and fold it 3 times through the middle. How many small triangles will you be able to count when you unfold it? Guess first. Then try it.

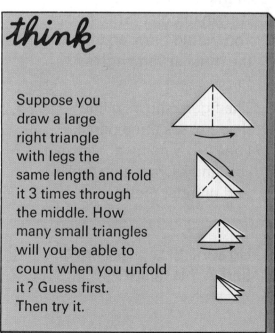

197

● *Let's explore angles and parallel lines.*

Investigating the Ideas

When a line crosses two **parallel lines**, eight angles are formed.

One angle is shown in red.

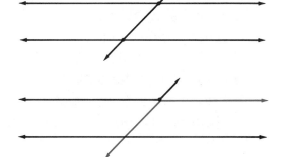

 Can you draw a figure like this and use your crayons to color another angle?

Discussing the Ideas

1. In the figure, angle 1 is shown in color. Show how you would draw and number the other seven angles.

2. The figure shows a line that crosses three parallel lines. How many angles can you find? Draw a picture and number them.

3. How many angles can you find in this figure?

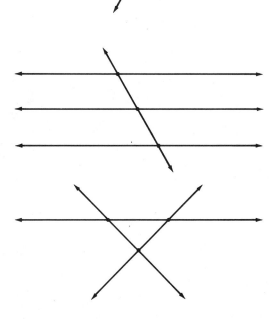

198

Using the Ideas

1. Draw a line that crosses two parallel lines.
   Number the eight angles that are formed as shown below.
   Color the insides of angles 1 and 2 and cut them out.

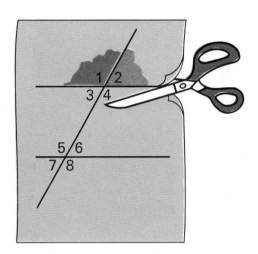
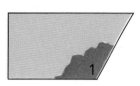

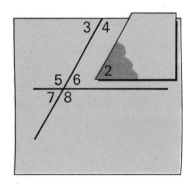

Which of the other angles
(3, 4, 5, 6, 7, or 8) are the
same size as angle 2? Use
angle 2 to help you find out.

2. Which of the other angles are
   the same size as angle 1?
   Use angle 1 to help you find out.

3. Draw a pair of lines that
   cross each other. Letter
   the angles as in the figure.
   A Which angle is the same
      size as angle A?
   B Which angle is the same
      size as angle B?

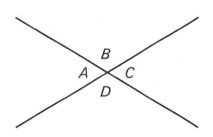

• Can you make some quadrilaterals?

Investigating the Ideas

These strips help you find the corners of a 4-sided figure.

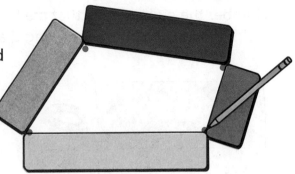

 Can you make a different 4-sided figure by using this set of strips? | Record your figure by marking points at the corners and connecting them.

Discussing the Ideas

1. A closed figure like this one is called a quadrilateral.
   A. A quadrilateral has __?__ line segments.
   B. Can you name some objects that are in the shape of quadrilaterals?

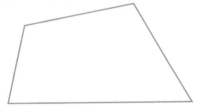

A quadrilateral

2. How many vertices does a quadrilateral have?

3. Can you mark 4 points on the chalkboard so that a quadrilateral is not formed when they are connected?

4. Can you find 4 of your strips that do not form a quadrilateral?

Using the Ideas

1. **A** Mark 4 points on your paper, as in the figure. Be sure that no 3 of them are in a line. Now use your ruler to connect the points like this. When you finish, you will have a quadrilateral.
   B Draw one segment that divides your quadrilateral into two triangles.

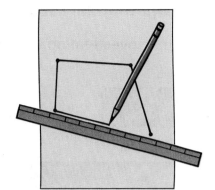

2. Draw a triangle. Draw one segment on this figure that divides the triangle into a smaller triangle and a quadrilateral.

3. Draw a quadrilateral. Mark a point outside the quadrilateral. From this point, draw a path that crosses the quadrilateral 4 times.
   A Where will your pencil point be then, inside or outside?
   B If the path crosses 9 times, where will it end, inside or outside?

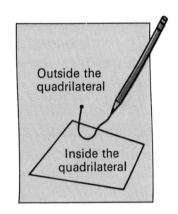

★ 4. Study the chart. Then draw and name 5 different quadrilaterals.

We see a quadrilateral	We label some points	We write a name for it	We say
(trapezoid shape)	A, B, D, C labeled	ABCD	"quadrilateral ABCD"

● *Can you name some special quadrilaterals?*

Investigating the Ideas

For this investigation, use these strips.

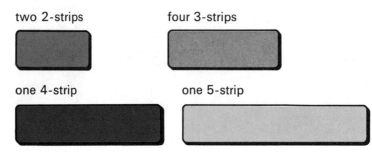

| Can you make one of the quadrilaterals described below? | List the strips you used. |

- A All four sides have the same length.
- B The two longer sides have the same length and the two shorter sides have the same length.
- C Two sides have different lengths and are parallel.
- D No two sides have the same length.

Discussing the Ideas

Special Quadrilaterals

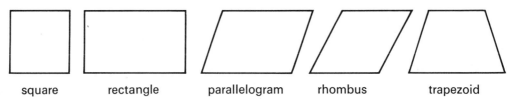

square   rectangle   parallelogram   rhombus   trapezoid

1. Which of the special quadrilaterals shown did you make in the Investigation?

2. In what ways are squares and rectangles alike? In what way are they different? Compare other figures in this way.

Using the Ideas

1. **A** What kind of figure is suggested by a sheet of tablet paper?
   B Use the method suggested in the picture to cut a square from tablet paper.

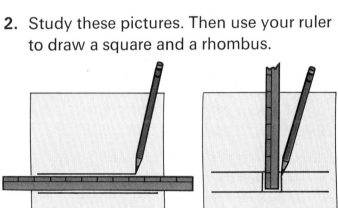

2. Study these pictures. Then use your ruler to draw a square and a rhombus.

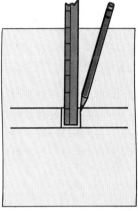
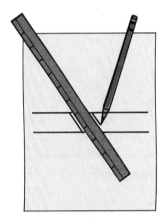

Draw a line along each side of your ruler.

Draw like this to form a **square**.

Draw like this to form a **rhombus**.

Can you use your ruler like this to draw a rectangle and a parallelogram?

★ 3. How much of a circle can you fill with the corners of a quadrilateral?

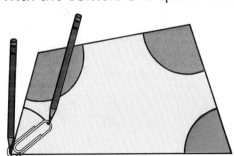

203

● *Let's explore parallelograms.*

Investigating the Ideas

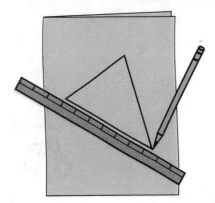

Draw a triangle on a piece of folded paper.

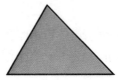 

Cut the folded paper to get two triangles that are just alike. Color them different colors.

 How many different quadrilaterals can you make by placing sides of your triangles together?

Discussing the Ideas

1. How many parallelograms did you find in the Investigation?

2. Jack made these two shapes with his triangles. Which one is a parallelogram?

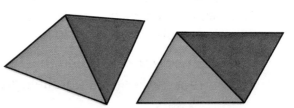

3. Can you cut out two triangles that are just alike and form a rectangle?

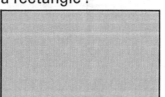

4. Could two triangles that are just alike form this shape?

Using the Ideas

1. Draw a parallelogram and cut it out. Then cut it into two triangles. Do they fit exactly upon each other?

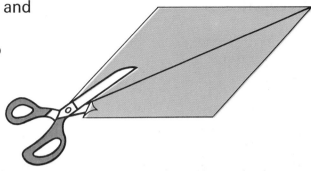

2. Draw a large 4-sided figure on your paper.

   A Color it and then cut it out.

   B Find the midpoint of each side by folding.

   C Connect the midpoints with segments as in the figure.

   D What kind of a figure do you think you have?

   E Do you think this would work if you started over with a different 4-sided figure?

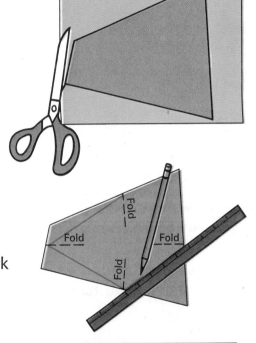

think

How many pennies does it take to make a stack as high as a penny standing on edge? 5? 6? 7? 8? 9? 10? 11? 13? 15? Guess. Then check your guess.

● *Let's explore some polygons.*

Investigating the Ideas

You can separate any quadrilateral region into two triangular regions by cutting along a **diagonal** (corner to corner).

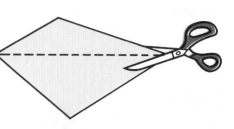

quadrilateral

Draw a **pentagon** (5-sided polygon) and a **hexagon** (6-sided polygon) and cut them out.

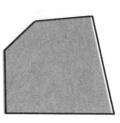 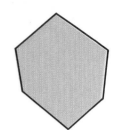

pentagon   hexagon

 Can you find how many triangles are formed by cutting along the diagonals from one corner of a pentagon? of a hexagon?

Discussing the Ideas

1. Can you explain how to complete the table?

Number of sides of polygon	Number of triangles formed
3	1
4	2
5						
6						
7						

2. 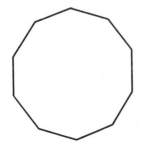 This polygon has ten sides and is called a **decagon**. Into how many triangles can it be separated by cutting along the diagonals from one corner?

Using the Ideas

1. **Regular polygons** have all sides and all angles the same. Copy and complete the table for the regular polygons shown.

Name of polygon	Number of sides	Number of angles	Any sides parallel?
Regular triangle (equilateral)			
Regular quadrilateral (square)			
Regular pentagon			
Regular hexagon			

2. Draw a **quadrilateral** on your paper.
    A Draw all the **diagonals**.
    B How many diagonals are there?
    C How many are there from each vertex?

3. Draw a **pentagon** on your paper. Choose one vertex and label it **A**.
    A How many diagonals can you draw from **A**?
    B Can you draw the same number of diagonals from any vertex?

★ 4. Give the total number of diagonals that can be drawn for each figure.
    A pentagon
    B hexagon
    C a 10-sided polygon

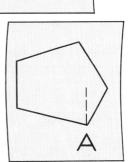

207

● *What is a simple closed curve?*

Investigating the Ideas

Here are four ways to draw a circle.

  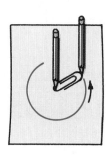 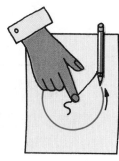

Drawing around a cup or a can   Using a compass   Using a paper clip   Using a string

 Can you use each of these methods to make a circle? Which method was easiest for you?

Discussing the Ideas

1. Each of these is a **simple closed curve**.

   None of these is a **simple closed curve**.

   Which of these are **simple closed curves**?

2. How can you tell whether or not a figure is a **simple closed curve**?

Using the Ideas

1.

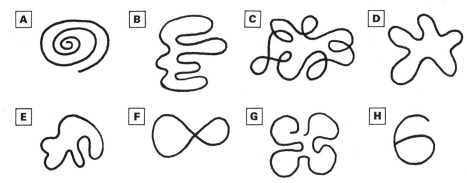

A Which curves above are closed?
B Which curves above are simple closed curves?
C Which curves above are closed but not simple?

2. Draw a simple closed curve. Mark a point inside the curve.
   A Start at the point and draw a path that crosses the curve. Where is your pencil point now, inside or outside?
   B If you cross 5 times in all, where are you? 6 times?

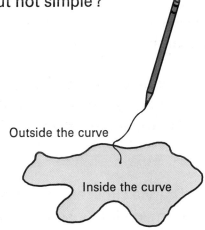

3. A Draw a circle and cut it out.
   B Find the **center** of the circle like this:
      ▲ Fold the circle as shown here.
      ▲ Open it and fold it again as shown but in a different place.
   C Where is the center of the circle?

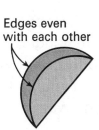

Edges even with each other

209

• *What is a symmetrical figure?*

Investigating the Ideas

This square is folded so that one half exactly matches the other half.

 How many different ways can you fold a square so that one half exactly matches the other?

Discussing the Ideas

1. A square is **symmetrical** because you can fold it so that one half exactly matches the other half. Are there other symmetrical figures?

2. Draw a circle and cut it out. Can you fold it so that the halves match? Can you do this in more than one way?

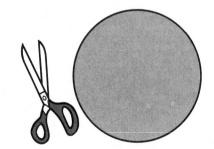

3. A sheet of tablet paper forms a rectangle. Can you fold it to form matching halves? Can you do this in more than one way?

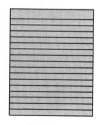

Using the Ideas

1. Here is a way to make a symmetrical figure.

   Fold a piece of paper.   Make a cut that starts and ends on the fold.   Unfold the piece you cut out. It will be symmetrical.

   Make cuts so that the unfolded shape will look like
   - A a rectangle.
   - B a leaf.
   - C a triangle.
   - D a square.
   - E a house.
   - F a pumpkin.
   - G a rocket.
   - H a hexagon.

2. Draw a picture to show what each cut-out piece below will look like when unfolded.

   A 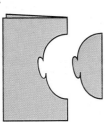   B 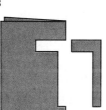   C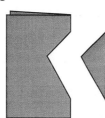

3. Write **S** (symmetrical) or **NS** (not symmetrical) for each figure. If it is symmetrical, think about how you would fold it to make the halves match.

   A   B   C   D

Reviewing the Ideas

1. Give the letter of the figure that matches each name.
   A square
   B parallelogram
   C rectangle
   D trapezoid

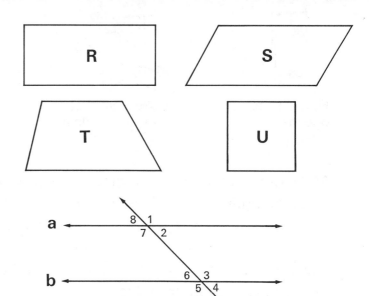

2. Lines **a** and **b** are parallel. Name all of the angles in the figure that are the same size as angle 2.

3. Draw a quadrilateral on your paper. Draw the diagonals of the quadrilateral.

4. Which of these shapes are symmetrical?

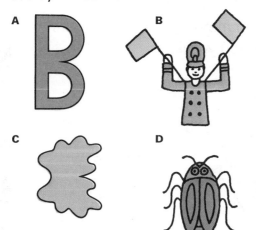

think

Fold a piece of paper twice and cut off the corner as shown. Guess what the cut-off piece will look like when unfolded and draw a picture. Then unfold it and check your guess.

Keeping in Touch with   Measurement   Subtraction
Place value   Fractions
Addition

1. Measure each object to the nearest ½ inch.

   A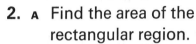

   B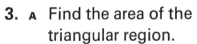

2. A Find the area of the rectangular region.
   B What is the area of half the rectangle?
   C What is the area of ¼ the rectangle?

3. A Find the area of the triangular region.
   B What is the area of half this region?

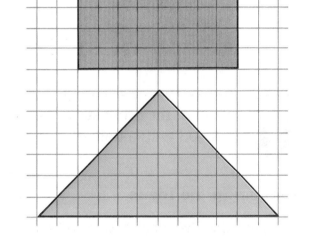

4. Solve the equations.
   A $56 = 40 + n$ C $41 = n + 11$ E $84 = n + 14$
   B $72 = 60 + n$ D $90 = 80 + n$ F $76 = n + 16$

5. Find the sums and differences.

   A $\;\;24$ B $\;\;78$ C $\;\;35$ D $\;\;73$ E $\;\;84$
   $\;\;+31$ $\;\;-46$ $\;\;+48$ $\;\;-26$ $\;\;+78$

   F $\;121$ G $\;362$ H $\;284$ I $\;375$ J $\;721$
   $\;\;-75$ $\;+475$ $\;-166$ $\;+468$ $\;-145$

   You are invited to explore   **ACTIVITY CARD 10** Page 314

9 Number Theory

● *What are odd and even numbers?*

Investigating the Ideas

You can match the 6-strip with a "train" of 2-strips.

 Which of the other strips can be matched with a train of 2-strips?

Discussing the Ideas

The numbers whose strips can be matched with a train of 2-strips are called **even numbers**.
Zero is also an even number.
The other numbers are called **odd numbers**.

1. A Name the even numbers less than 50.
   B Name the odd numbers between 50 and 100.

2. Every odd-numbered strip can be matched with a train of 2-strips and how many extra 1-strips?

3. Can you think of an easy way to decide whether a number is even or odd?

214

Using the Ideas

1. Give the function rules and the missing numbers.

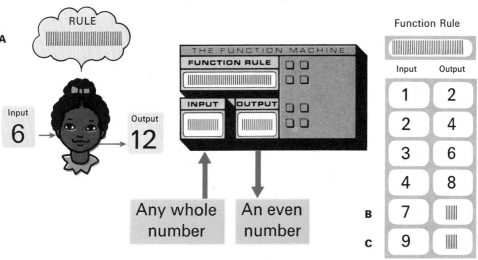

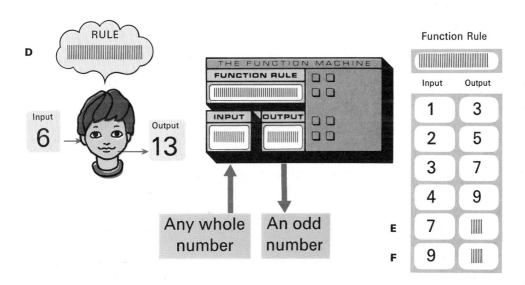

2. Complete a table like this one for these numbers: 34, 35, 47, 48, 60, 61.

Number	Equation	Odd or Even
24	24 = **12** + **12**	Even
25	25 = **12** + **13**	Odd

● *What are some patterns of odd and even numbers?*

Investigating the Ideas

	Input	Output
	19	9
	20	0
	76	6
A	45	
B	960	
C	871	
D	408	

? Can you find Jean's function rule and use it to complete the table?

Discussing the Ideas

1. Can you tell whether a number is even or odd if you know only the last digit of the numeral?

2. Study the examples. Then complete the exercises.
   Examples: 57 ends with 7.   57 is an odd number.
   　　　　　　86 ends with 6.   86 is an even number.
   A 34 ends with ▓. Is 34 an even or an odd number?
   B 43 ends with ▓. Is 43 an even or an odd number?
   C 30 ends with ▓. Is 30 an even or an odd number?
   D 138 ends with ▓. Is 138 an even or an odd number?
   E 469 ends with ▓. Is 469 an even or an odd number?

3. Is the number for this 3-digit numeral odd or even?

4. Answer "even" or "odd."
   A Each _?_ number ends with 0, 2, 4, 6, or 8.
   B Each _?_ number ends with 1, 3, 5, 7, or 9.

Using the Ideas

1. Find the sums or products. Then tell whether "even" or "odd" should go in the blank.

   A   6   16   60   78   D   4   2   6   8
      +8   +28   +78   +54   ×6   ×6   ×8   ×2

   The sum of two even numbers is an __?__ number.   The product of two even numbers is an __?__ number.

   B   5   15   37   65   E   7   3   1   7
      +7   +1   +45   +87   ×3   ×5   ×7   ×5

   The sum of two odd numbers is an __?__ number.   The product of two odd numbers is an __?__ number.

   C   6   47   38   57   F   6   2   3   6
      +7   +6   +11   +38   ×3   ×5   ×8   ×7

   The sum of an even and an odd number is an __?__ number.   The product of an even and an odd number is an __?__ number.

2. Answer "even" or "odd."
   - A The sum of an even number and 1 is an __?__ number.
   - B The sum of an odd number and 1 is an __?__ number.
   - C The product of an odd number and 1 is an __?__ number.
   - D No __?__ number is less than 1.
   - E Every __?__ number is greater than 0.
   - F There are two __?__ numbers less than 3.
   - G There is only one __?__ number less than 3.
   - H The product of 0 and an odd number is an __?__ number.
   - I The sum of an even number and 0 is an __?__ number.
   - J The product of an even number and 0 is an __?__ number.
   - K The sum of two odd numbers is an __?__ number.
   - L The product of two odd numbers is an __?__ number.
   - M The sum of an even and an odd number is an __?__ number.
   - N The product of an even and an odd number is an __?__ number.

● Can you list multiples of a number?

Discussing the Ideas

×	0	1	2	3	4	5	6	7	8	9	10	11	12	13	14
A 2	0	2	4	6	8	10	12	14	16	18	20	22	24	26	
B 3	0	3	6	9	12	15	18	21	24	27	30	33	36	39	
C 4	0	4	8	12	16	20	24	28	32	36	40	44	48	52	

The numbers in row **A**, {0, 2, 4, 6, . . .}, are **multiples** of 2.

The numbers in row **B**, {0, 3, 6, 9, . . .}, are **multiples** of 3.

The numbers in row **C**, {0, 4, 8, 12, . . .}, are **multiples** of 4.

1. Give a multiple of 2 not shown in the table. How does the diagram above suggest another way to describe the even numbers?

2. Give a multiple of 3 not shown in the table. Use the table to count by threes to 39.

3. Find a number in the table (other than 0) that is a multiple of 3 and a multiple of 4. Find another such number.

4. Why would 18 be in the list of multiples of 6?

★ 5. The rectangle contains **only** multiples of 2. The circle contains **only** multiples of 3. Can you explain how to find some more numbers for the shaded space?

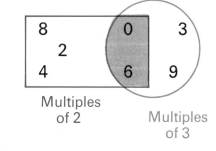

Multiples of 2   Multiples of 3

Using the Ideas

1. List the multiples
   - A of 5 up to 50.
   - B of 6 up to 60.
   - C of 7 up to 70.
   - D of 8 up to 80.
   - E of 9 up to 90.
   - F of 10 up to 100.

think

Of the numbers with 2 digits,
I'm the smallest with this fate.
I'm a multiple of six
And a multiple of eight.

WHO AM I?

2. Find the missing numbers.
   - A 8 is a multiple of 2.
     8 is also a multiple of ▦.
   - B 6 is a multiple of both ▦ and ▦.
   - C Since 3 × 4 = 12, 12 is a multiple of both 3 and ▦.
   - D Since 5 × 6 = 30, 30 is a multiple of both ▦ and 6.
   - E Since 7 × 8 = 56, 56 is a multiple of both ▦ and ▦.

3.
   - A In the table on page 218, find a number (other than 0) that is a multiple of 2, 3, and 4.
   - B Find another multiple of 2, 3, and 4.
   - C Find a number less than 50 that is a multiple of 2, 4, 5, 8, and 10.

★ 4. Write the numbers 1 through 100 in rows as shown. Circle the multiples of 5. What pattern do you see? Mark the multiples of another number. Is there a pattern? Mark multiples of other numbers to show as many patterns as you can.

```
 1  2  3  4  5  6  7  8  9 10
11 12 13 14 15 16 17 18 19 20
21 22 23 24 25 26 27 28 29 30
31  . . .
 .
 .
 .
```

● Can you find the factors of a number?

Investigating the Ideas

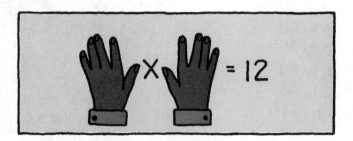

 Can you record all the different equations that might be hidden?

Discussing the Ideas

1. Study this diagram.

    3 × 4 = 12   Since the product of 4 and 3 is 12, we say
    —— 4 **is a factor of 12**   and
    —— 3 **is a factor of 12**.

    What are some other factors of 12?

2. Kevin is covering up two factors of 20.

    A Can you be certain about what these factors are?
    B Give a pair of numbers that Kevin might be hiding.
    C Write three equations, using different factors that might be on the board.
    D List six numbers that are factors of 20.
    E Is 7 a factor of 20? Why?

Using the Ideas

1. **A** Write a third equation to show other factors of 18.
   B List six different factors of 18.
   C Is 4 a factor of 18?

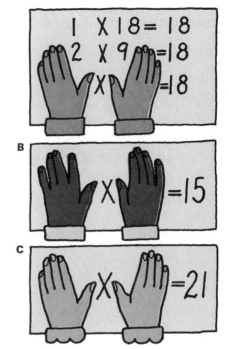

2. For each exercise, write as many equations as you need to show all the factors of the product.

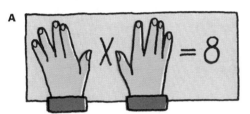

3. List all the factors of each number. Use exercise 2 if you need help.
   A 8   **B** 15   **C** 21

4. **A** Which of the numbers 3, 4, 5 is a factor of 16?
   B Which equation can you solve?
   1. $16 \div 3 = n$
   2. $16 \div 4 = n$
   3. $16 \div 5 = n$

★ 5. Which of the numbers 2, 3, 4, 5, 6 are factors of:
   A 15? (Answer: 3 and 5)
   B 10?   **F** 30?
   C 20?   **G** 31?
   D 24?   **H** 32?
   E 28?   **I** 60?

I'm proud to be a factor of
Both twelve and forty-two.
No larger number makes this claim.
You need no other clue.
WHO AM I?

• Which numbers are prime?

Investigating the Ideas

The figures show which numbers from 2 through 8 can be shown as rectangles using sets of squares. No single strings of squares are allowed!

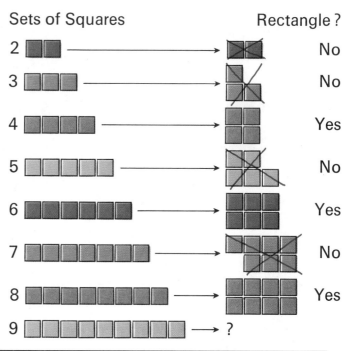

 Can you use sets of squares to find which numbers from 9 through 20 will form rectangles?

Record your findings on graph paper.

Discussing the Ideas

1. The numbers greater than 1 that do not make "rectangles" are called **prime numbers**.
    A Which numbers from 2 through 8 are prime numbers?
    B What prime numbers did you find from 9 through 20?

2. Which numbers between 20 and 30 do you think are prime numbers?

3. The number 39 is the product of 3 and 13. How could you use this fact to convince someone that 39 is not a prime number?

Using the Ideas

1. The product in each exercise is a prime number. Write as many equations as you need to show all the factors of the product.

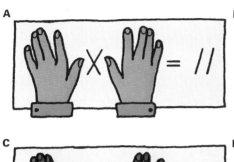
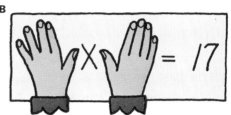

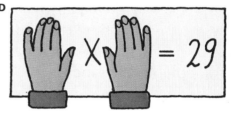

2. **A** How many factors did each prime number in exercise 1 have?
   B Can you find a prime number with more than two factors?

3. Write an equation for each exercise.
   Example: I know 20 is **not prime** because $4 \times 5 = 20$.
   A I know 32 is **not prime** because ▦ × ▦ = 32.
   B I know 33 is **not prime** because ▦ × ▦ = 33.

4. List the prime numbers between 40 and 50.

Now I'm a strange prime number.
No others are like me.
All my cousins are so odd,
As odd, as odd can be.

WHO AM I?

More practice, page A–27, Set 37

Reviewing the Ideas

1. Which numbers are even and which are odd?
   A 6   B 9   C 5   D 12   E 32   F 31   G 146   H 283

2. Find the missing numbers.
   A 14 is a multiple of 2 because $2 \times 7 = n$.
   B 12 is a multiple of 3 because $n \times 4 = 12$.
   C Since $4 \times 6 = 24$, we call ▒ a multiple of 4.
   D Since $n \times 12 = 36$, we call 36 a multiple of 3.
   E Since $5 \times 6 = 30$, we know that ▒ is a multiple of 5.
   F 45 is a multiple of 5 since $5 \times 9 = n$.

3. Find the missing numbers.
   A ▒ and ▒ are factors of 18 because $3 \times 6 = 18$.
   B 2 and 9 are factors of ▒ because $2 \times n = 18$.
   C Since $4 \times 5 = 20$, ▒ and ▒ are factors of 20.
   D Since $2 \times 3 \times 4 = 24$, 2, 3, and 4 are factors of ▒.
   E Since $21 \div 3 = 7$, ▒ and ▒ are factors of 21.

4. List the factors of the following numbers.
   A 6   B 5   C 9   D 10   E 11   F 12   G 20   H 29

★ 5. Give the digits.
   A If a number is a multiple of 2, then it ends in one of the digits ▒, ▒, ▒, ▒, or ▒.
   B If a number is a multiple of 5, then it ends in one of the digits ▒ or ▒.
   C The multiples of 3 may end in any one of the digits 0 to 9. Show this by listing the first ten multiples of 3.
   D What other numbers between 1 and 10 have multiples that may end in any one of the digits 0 to 9?

Keeping in Touch with

Addition
Subtraction
Measurement

Place value
Inequalities

1. Solve the equations.
   A $84 = 80 + n$ B $384 = 300 + 80 + n$ C $58 = n + 8$
   D $458 = n + 50 + 8$ E $763 = 700 + n + 3$

2. Find the sums and differences.

   A $24 + 35$ B $78 - 26$ C $42 + 30$ D $79 - 20$ E $65 + 18$ F $94 - 56$

   G $38 + 27$ H $54 - 19$ I $76 + 88$ J $134 - 75$ K $84 + 16$ L $100 - 23$

3. Solve the equations.
   A $7 + n = 15$ B $16 - 9 = n$ C $13 - 7 = n$
   D $n + 5 = 12$ E $10 - 4 = n$ F $6 + n = 14$

4. Give the sign $>$, $<$, or $=$ for each ⬤.
   A 482 ⬤ 472 B 6286 ⬤ 6296 C $50 + 8$ ⬤ $50 - 8$
   D $70 - 2$ ⬤ $70 - 3$ E $70 + 4$ ⬤ $70 + 5$ F $80 + 0$ ⬤ $80 - 0$

5. Find the area of each region.
6. Find the volume of each region.

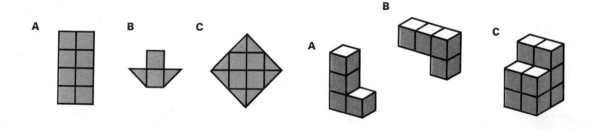

You are invited to explore

ACTIVITY CARD 11
Page 314

10 Multiplying

● *Is there an easy rule for multiplying by 10 and 100?*

Investigating the Ideas

 Can you give a rule for multiplying by 10? by 100?

Discussing the Ideas

1. A How many sets of 10 are in figure A?
   B Write the numeral for 6 tens.
   C Solve: $6 \times 10 = n$

   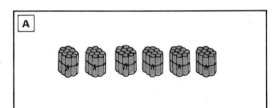

2. A How many sets of 100 are in figure B?
   B Write the numeral for 3 hundreds.
   C Solve: $3 \times 100 = n$

   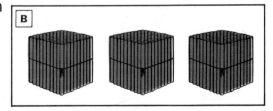

3. A Explain your rule for multiplying a 1-digit number by 10.
   B Explain your rule for multiplying a 1-digit number by 100.

Using the Ideas

1.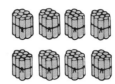
   A How many tens?
   B Write the numeral for this number of tens.
   C Solve: $8 \times 10 = n$

2.
   A How many hundreds?
   B Write the numeral for this number of hundreds.
   C Solve: $2 \times 100 = n$

3. Find the products.
   A 5×10 E 8×100 I 9×100 M 3×100
   B 7×100 F 9×10 J 7×10 N 6×10
   C 2×10 G 4×10 K 8×10 O 1×100
   D 4×100 H 5×100 L 2×100 P 6×100

4. Solve the equations.
   A $6 \times 10 = n$ F $n \times 10 = 50$ K $4 \times n = 40$
   B $3 \times 10 = n$ G $n \times 100 = 300$ L $6 \times n = 600$
   C $4 \times 100 = n$ H $n \times 10 = 40$ M $7 \times n = 70$
   D $7 \times 10 = n$ I $n \times 100 = 600$ N $3 \times n = 300$
   E $9 \times 100 = n$ J $n \times 100 = 800$ O $2 \times n = 200$

think

Study the pattern.
Then copy the equations,
giving the missing numbers.

$1 \times 9 = 10 - 1$
$2 \times 9 = 20 - 2$
$3 \times 9 = 30 - 3$
$4 \times 9 = 40 - 4$
$5 \times 9 = \blacksquare - \blacksquare$
$8 \times 9 = \blacksquare - \blacksquare$
$13 \times 9 = \blacksquare - \blacksquare$

● *Does the "10 rule" work for products like 23 × 10?*

Investigating the Ideas

If your white strip ( ) is 1, how long is this train?

If your white strip were 10, how long would this train be?

Can you show 240 with your strips if you think of your white strip as 10?

Discussing the Ideas

1. Explain how you can use strips to help you think of 14 × 10.

2. A Explain how Jack is thinking about 23 × 10.
   B Write the numeral for 20 tens.
   C Write the numeral for 3 tens.
   D Solve: 23 × 10

3. Jill knows a shortcut for multiplying by ten. She made a chart to help the other children discover her rule.
   A What is Jill's rule for multiplying by 10?
   B What is a simple rule for multiplying by 100?

228

Using the Ideas

1. Find the value in cents for each set of dimes.

2. Give the value in cents for each coin collection.
   - A 5 dimes
   - B 10 dimes
   - C 20 dimes
   - D 15 dimes
   - E 27 dimes
   - F 40 dimes
   - G 52 dimes
   - H 91 dimes
   - I 56 dimes
   - J 48 dimes

3. Solve the equations.
   - A $(40 \times 10) + (3 \times 10) = n$
     $43 \times 10 = n$
   - B $(70 \times 10) + (2 \times 10) = n$
     $72 \times 10 = n$
   - C $(30 \times 10) + (1 \times 10) = n$
     $31 \times 10 = n$
   - D $(60 \times 10) + (5 \times 10) = n$
     $65 \times 10 = n$
   - E $(20 \times 10) + (9 \times 10) = n$
     $29 \times 10 = n$
   - F $(50 \times 10) + (6 \times 10) = n$
     $56 \times 10 = n$

4. Find the products.
   - A $10 \times 10 = n$
   - B $27 \times 10 = n$
   - C $10 \times 43 = n$
   - D $50 \times 10 = n$
   - E $10 \times 19 = n$
   - F $96 \times 10 = n$
   - G $10 \times 73 = n$
   - H $10 \times 65 = n$
   - I $34 \times 10 = n$
   - J $80 \times 10 = n$
   - K $8 \times 100 = n$
   - L $52 \times 100 = n$

think

Now I am a number
You rarely can beat.
When I am a factor
I surely am neat.
Use the other factor.
Make zero the tail.
You'll see the product.
You really can't fail.

WHO AM I?

More practice, page A-28, Set 38

● Let's explore products like 3 × 40 and 3 × 400.

Discussing the Ideas

1.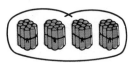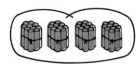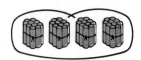

   A How many tens in each ring?
   B How many tens in all?
   C How many sticks in all?
   D 3 × 4 tens is how many tens?
   E Solve: 3 × 40 = **n**

2.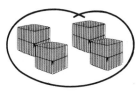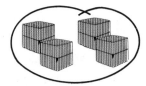

   A How many hundreds in each ring?
   B How many hundreds in all?
   C How many sticks in all?
   D 3 × 4 hundreds is how many hundreds?
   E Solve: 3 × 400 = **n**

3. A Explain how Fran is thinking about the product 5 × 30.
   B Write the numeral for 15 tens.
   C Solve: 5 × 30 = **n**
   D How would Fran think about 2 × 70? 4 × 60? 7 × 30?

Fran

Using the Ideas

1. Give the missing numbers.

4×20 5×30 6×40

A $4 \times \text{\#\#\#\#}$ tens
   $\text{\#\#\#\#}$ tens in all
   $4 \times 20 = n$

B $5 \times \text{\#\#\#\#}$ tens
   $\text{\#\#\#\#}$ tens in all
   $5 \times 30 = n$

C $6 \times \text{\#\#\#\#}$ tens
   $\text{\#\#\#\#}$ tens in all
   $6 \times 40 = n$

2. Solve the equations.

A $(4 \times 2) \times 10 = n$
   $4 \times (2 \times 10) = n$
   $4 \times 20 = n$

B $(5 \times 3) \times 10 = n$
   $5 \times (3 \times 10) = n$
   $5 \times 30 = n$

C $(6 \times 4) \times 10 = n$
   $6 \times (4 \times 10) = n$
   $6 \times 40 = n$

D $(7 \times 3) \times 10 = n$
   $7 \times (3 \times 10) = n$
   $7 \times 30 = n$

3. Find the products.

A 4×6
   4×60
   4×600

B 6×3
   6×30
   6×300

C 3×7
   3×70
   3×700

D 5×7
   5×70
   5×700

E 3×9
   3×90
   3×900

F 6×6
   6×60
   6×600

G 7×4
   7×40
   7×400

H 5×8
   5×80
   5×800

4. Find the products.

A 3×70
B 3×700
C 4×30
D 4×300
E 2×80
F 2×800
G 6×10
H 6×100
I 7×20
J 7×200
K 8×40
L 3×80
M 2×90
N 9×30
O 5×50
P 8×400
Q 2×900
R 9×300
S 5×500
T 4×800

think

If it's products you are after,
I will lend a helping hand.
Just use me over and over,
On the answer you will land.

WHO AM I?

More practice, page A-29, Set 39

Pike
10 miles per hour

Runner
20 miles per hour

Sailboat
30 miles per hour

Race horse
40 miles per hour

Deer
50 miles per hour

Homing pigeon
60 miles per hour

Cheetah
70 miles per hour

Story Problems

1. A helicopter can fly about 3 times as fast as a homing pigeon. How fast can the helicopter fly?

2. Racing bikes can be made to go 2 times as fast as a deer. How fast can racing bikes go?

3. A jet speedboat goes 9 times as fast as a sailboat. How fast does a jet speedboat go?

4. What object can go 10 times as fast as a pike can swim?

5. A racing car runs for 9 hours. A jet flies for 7 hours. Which goes farther? How much farther?

6. A propeller plane can fly 5 times as fast as a vulture. How fast can a propeller plane fly?

7. A golden eagle can fly about 6 times as fast as a runner can run. How fast can the eagle fly?

8. A bullet goes how many times as fast as a pitched baseball?

9. If a cheetah could run 6 times as fast as usual, could it catch a racing car? How fast would the cheetah be running?

10. A man flies in a jet plane from Washington, D.C., to San Francisco, California. It takes him 4 hours for the trip. About how far is it between the two cities?

11. A boy rode for 7 hours on a monorail train. How far did he ride?

★12. A homing pigeon was taken 300 miles from home. About how many hours would it take the pigeon to fly back home?

★13. A racing car can go how many times as fast as a deer?

★14. How much faster than its usual rate would a vulture have to fly to pass a jet plane?

★15. A satellite circles the earth for 5 hours. How far does it travel?

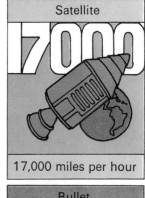
Satellite
17,000 miles per hour

Bullet
1400 miles per hour

Jet plane
600 miles per hour

Monorail train
80 miles per hour

Vulture
90 miles per hour

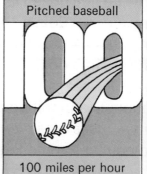
Pitched baseball
100 miles per hour

Racing car
400 miles per hour

Solving Multiplication-Addition Problems

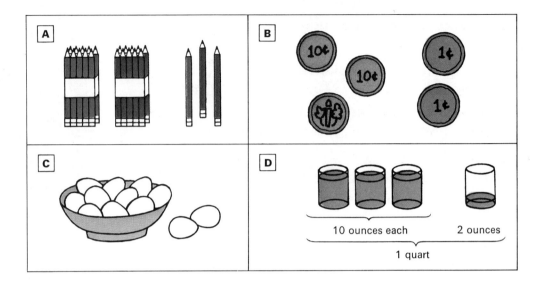

1. A How many pencils do you see in figure A?
   B If Juan has 2 times as many pencils, how many does he have?
   C Suppose Rosa has 3 times as many pencils as in A. How many pencils does she have?

2. A What is the value in cents of the money pictured in figure B?
   B If Tom has twice that much money, how much does he have?
   C Suppose a toy cost 3 times as much as the amount pictured in B. How much does the toy cost?

3. A One dozen is 10 and 2. (See figure C.) How many are in a dozen?
   B How many are in 3 dozen?
   C How many are in 4 dozen?
   D How many are in 5 dozen?

4. A Look at figure D and tell how many ounces are in a quart.
   B How many ounces are in 2 quarts?
   C How many ounces are in 3 quarts?
   D How many ounces are in 5 quarts?

Solving Story Problems

MAILING PACKAGES

Package	Stamps needed
A	30¢, 1¢, 1¢
B	20¢, 1¢
C	10¢, 5¢
D	30¢, 5¢, 1¢, 1¢

1. **A** How much did it cost to mail package A?
   B How much would it cost to mail 3 packages like this one?

2. **A** How much did it cost to mail package B?
   B How much would it cost to mail 4 packages like this one?

3. **A** How much did it cost to mail package C?
   B How much would it cost for 3 such packages?

4. **A** How much did it cost to mail package D?
   ★ **B** How much would it cost to mail 5 such packages?

235

● *Let's explore the multiplication-addition principle again.*

Investigating the Ideas

Cut a 4-by-8 rectangle from graph paper. Color each row of 8 a different color.

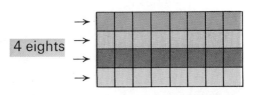

Another way to think about 4 eights is shown by the cut. Now you have 4 fives and 4 threes.
$4 \times 8 = (4 \times 5) + (4 \times 3)$

 Can you show how to think about 4 eights in a different way by cutting another 4-by-8 rectangle?

Discussing the Ideas

1. Study the figure below. Give the missing number.

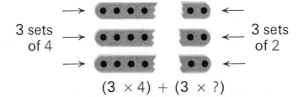

2. Carol's hands are covering two numbers.
   A Could the numbers be 6 and 2?
   B Could the numbers be 3 and 2?
   C Write 5 different equations to show what pairs of numbers Carol might be hiding.
   D Repeat part C for $5 \times 9 = (5 \times $ 🖐 $) + (5 \times $ 🖐 $)$.

Carol

Using the Ideas

1. Give the missing word.

A (4 fives and 4 __?__)
   4 × 7

B (7 threes and 7 __?__)
   7 × 5

C (5 fives and 5 __?__)
   5 × 7

D (7 sixes and 7 __?__)
   7 × 8

E (3 fives and 3 __?__)
   3 × 9

F (6 fours and 6 __?__)
   6 × 7

G (4 tens and 4 __?__)
   4 × 12

H (2 twenties and 2 __?__)
   2 × 26

2. Find the missing number.
- A 8 × 6 = (8 × 3) + (8 × n)
- B 4 × 7 = (4 × 5) + (4 × n)
- C 3 × 8 = (3 × 4) + (3 × n)
- D 7 × 7 = (7 × 2) + (7 × n)
- E 9 × 3 = (9 × n) + (9 × 2)
- F 6 × 12 = (6 × 10) + (6 × n)
- G 5 × 18 = (5 × 10) + (5 × n)
- H 3 × 23 = (3 × n) + (3 × 3)

think

Subtract 24 from me,
Or take me from 42.
Your answer is the same,
Whichever one you do.

WHO AM I?

● How can you use the multiplication-addition principle?

Discussing the Ideas

1. Explain how Sue is thinking about the product 3 × 12.

2. Write the numeral for 3 tens.

3. Write the numeral for 3 twos.

4. Give the product for 3 × 12.

5. How would Sue think about 2 × 13?

6. Find the product for 2 × 13.

7. Give the missing numbers.

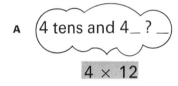
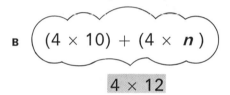

8. Solve the equations.
   A 4 × 12 = (4 × 10) + (4 × *n*)
   B 2 × 13 = (2 × 10) + (2 × *n*)
   C 3 × 11 = (3 × 10) + (3 × *n*)
   D 5 × 11 = (5 × 10) + (5 × *n*)
   E 3 × 23 = (3 × *n*) + (3 × 3)
   F 5 × 14 = (5 × *n*) + (5 × 4)
   G 4 × 24 = (4 × *n*) + (4 × 4)
   H 3 × 27 = (3 × *n*) + (3 × 7)

9. A What is 4 × 10?
    B What is 4 × 5?
    C What is (4 × 10) + (4 × 5)?
    D What is 4 × 15?

10. A What is 3 × 10?
    B What is 3 × 4?
    C What is (3 × 10) + (3 × 4)?
    D What is 3 × 14?

Using the Ideas

Find the products and sums.

1.
   A. $3 \times 10 = n$
   B. $3 \times 2 = n$
   C. $(3 \times 10) + (3 \times 2) = n$
   D. $3 \times 12 = n$

2.
   A. $4 \times 10 = n$
   B. $4 \times 2 = n$
   C. $(4 \times 10) + (4 \times 2) = n$
   D. $4 \times 12 = n$

3.
   A. $2 \times 10 = n$
   B. $2 \times 3 = n$
   C. $(2 \times 10) + (2 \times 3) = n$
   D. $2 \times 13 = n$

4.
   A. $2 \times 30 = n$
   B. $2 \times 4 = n$
   C. $(2 \times 30) + (2 \times 4) = n$
   D. $2 \times 34 = n$

5.
   A. $2 \times 20 = n$
   B. $2 \times 3 = n$
   C. $(2 \times 20) + (2 \times 3) = n$
   D. $2 \times 23 = n$

6.
   A. $3 \times 20 = n$
   B. $3 \times 3 = n$
   C. $(3 \times 20) + (3 \times 3) = n$
   D. $3 \times 23 = n$

7.
   A. $6 \times 10 = n$
   B. $6 \times 3 = n$
   C. $(6 \times 10) + (6 \times 3) = n$
   D. $6 \times 13 = n$

8.
   A. $4 \times 20 = n$
   B. $4 \times 4 = n$
   C. $(4 \times 20) + (4 \times 4) = n$
   D. $4 \times 24 = n$

think

1-unit
Make 8 of these

2-unit
Make 5 of these

This is a game for two. The object is to cover the 10-unit strip exactly with the 1-unit and 2-unit pieces. Start at the left and take turns placing either a 1-unit or a 2-unit side by side until the 10-unit is exactly covered. The last one to put down a strip wins the game.

Start									

Try this game using an 11-unit strip. Try it with a 12-unit strip.

More practice, page A-29, Set 40

● *How can you find products like 3 × 54?*

Investigating the Ideas

You know that
3 × 54 = (3 × 50) + (3 × 4).

Let's use a function-machine table to help us find this product. What is the missing number in the table?

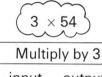

Multiply by 3	
input	output
4	12
50	150
54	

Can you make tables of your own to find these products?
1. 2 × 34 **2.** 4 × 36 **3.** 3 × 65

Discussing the Ideas

Let's look at a shorter way to write the work.

3 × 48

Multiply by 3	
8	24
40	120
48	144

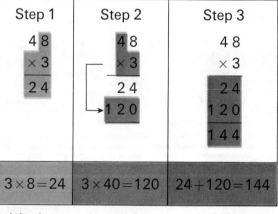

1. Which part of the table is like step 1?
2. Which is like step 2?
3. Which is like step 3?
4. Now try the same method with these.
   A 3 × 64   B 2 × 76

Using the Ideas

Find the products.

1. 14 ×5	2. 38 ×2	3. 24 ×3	4. 19 ×4	5. 47 ×4	6. 34 ×2
7. 67 ×3	8. 35 ×2	9. 36 ×5	10. 58 ×3	11. 75 ×4	12. 17 ×4
13. 18 ×3	14. 22 ×4	15. 94 ×3	16. 26 ×3	17. 29 ×5	18. 18 ×2
19. 31 ×8	20. 73 ×4	21. 19 ×3	22. 42 ×7	23. 27 ×3	24. 32 ×8

Short Stories

1. 12 dogs, 4 legs each. How many legs in all?

2. 14 ants, 6 legs per ant. How many legs?

3. 12 spiders, 8 legs per spider. How many legs?

4. 13 octopuses, 8 arms each. How many arms?

5. 14 crayfish, 10 legs each. How many legs?

6. 36 crickets, 6 legs each. How many legs?

7. 42 fish, no legs per fish. How many legs in all?

8. 4 centipedes, 36 legs each. How many legs?

More practice, page A-30, Set 41

● *Let's look at a shortcut for finding products.*

Discussing the Ideas

1. Explain each step of the long method.

2. Explain how step 1 of the shortcut is like step 1 of the long method.

3. Explain how step 2 of the shortcut puts together step 2 and step 3 of the long method.

4. Explain each step in this exercise.

    $$\begin{array}{r} 2 \\ 1\,8 \\ \times\ 3 \\ \hline 5\,4 \end{array}$$

Long Method		
Step 1	Step 2	Step 3
4 6 × 3 ───── 1 8	4 6 × 3 ───── 1 8 1 2 0	4 6 × 3 ───── 1 8 1 2 0 ───── 1 3 8
3 × 6 = 18	3 × 40 = 120	18 + 120 = 138

Shortcut	
Step 1	Step 2
1 4 6 × 3 ───── 8	1 4 6 × 3 ───── 1 3 8
3 × 6 = 18	3 × 40 = 120 120 + 10 = 130

5. Now try the shortcut with this one. 4 × 23

6. Carlos used the long method, but he made a mistake. Can you explain what he did wrong and find the correct product?

7. Beth Ann used the shortcut, but she also made a mistake. Can you explain what she did wrong?

Carlos
$$\begin{array}{r} 24 \\ \times 6 \\ \hline 24 \\ 1\,2 \\ \hline 36 \end{array}$$

Beth Ann
$$\begin{array}{r} 24 \\ \times 6 \\ \hline 124 \end{array}$$

242

Using the Ideas

In exercises 1 through 18, find the product.

1. 24 × 4	2. 15 × 3	3. 38 × 2	4. 24 × 6	5. 19 × 3	6. 47 × 3
7. 39 × 2	8. 21 × 4	9. 30 × 2	10. 63 × 4	11. 37 × 2	12. 37 × 3
13. 26 × 3	14. 54 × 5	15. 67 × 5	16. 75 × 6	17. 54 × 4	18. 68 × 4

Solving Story Problems

1. Mr. McCoy can drive his car about 17 miles on one gallon of gas. How far can he drive on 5 gallons of gas?

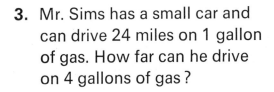

2. Mr. Ito's car goes only 14 miles on each gallon of gas. How far can he drive on 7 gallons of gas?

3. Mr. Sims has a small car and can drive 24 miles on 1 gallon of gas. How far can he drive on 4 gallons of gas?

4. Mr. Gomez can drive 33 miles on 2 gallons of gas. Mr. Perez can drive 16 miles on 1 gallon of gas. Who can drive farther

   A on 4 gallons of gas?
   B on 2 gallons of gas?
   C on 1 gallon of gas?
   D on 1 quart of gas?

Together, a doll and a dress cost $11. The doll cost $10 more than the dress. How much was the doll?

More practice, page A-31, Set 42

Building Multiplication Skills

1. Find the products.

   A. 3 × 6
   B. 7 × 4
   C. 3 × 9
   D. 5 × 8
   E. 9 × 8
   F. 7 × 7
   G. 3 × 7
   H. 8 × 3
   I. 4 × 9
   J. 6 × 8
   K. 9 × 9
   L. 8 × 7
   M. 6 × 4
   N. 4 × 8
   O. 9 × 5
   P. 5 × 7
   Q. 7 × 9
   R. 6 × 6
   S. 5 × 6
   T. 5 × 5
   U. 9 × 6
   V. 6 × 7
   W. 5 × 4
   X. 8 × 8

2. Find the products.

   A. 6 × 10
   B. 10 × 9
   C. 40 × 4
   D. 15 × 10
   E. 5 × 30
   F. 6 × 50
   G. 10 × 34
   H. 80 × 10
   I. 7 × 20
   J. 10 × 68
   K. 9 × 90
   L. 10 × 84
   M. 7 × 30
   N. 80 × 6
   O. 5 × 90
   P. 70 × 4
   Q. 3 × 90
   R. 2 × 60
   S. 70 × 10
   T. 85 × 10

3. Find the products.

   A. 32 × 3
   B. 43 × 2
   C. 44 × 2
   D. 17 × 2
   E. 28 × 3
   F. 19 × 4

4. Find the products.

   A. 47 × 4
   B. 56 × 5
   C. 87 × 2
   D. 27 × 3
   E. 56 × 4
   F. 65 × 6
   G. 73 × 6
   H. 65 × 5
   I. 76 × 3
   J. 74 × 8
   K. 67 × 7
   L. 89 × 4

think

Jack started with a number and multiplied it by 7. Then he added 75, subtracted 75 and divided by 7. Jack's answer was 19. What number did Jack start with?

Solving Story Problems TIME

1. There are 60 seconds in 1 minute. How many seconds are in 5 minutes?

2. There are 60 minutes in 1 hour. How many minutes are in 9 hours?

3. There are 24 hours in 1 day, and 7 days in a week. How many hours are in a week?

4. There are 12 months in 1 year. How many months are in 8 years?

5. There are about 52 weeks in 1 year. About how many weeks are in 6 years?

6. In each year, there are 7 months that have 31 days each. How many days in all are in these 7 months?

7. In each year, there are 4 months that have 30 days each. How many days in all are in these 4 months?

★ 8. How many days are in a year when February has 28 days? (Use your answers to exercises 6 and 7.)

★ 9. How many days are in a year when February has 29 days?

★ 10. How many seconds are in one hour?

★ 11. How many seconds are in one day?

S	M	T	W	T	F	S
1	2	3	4	5	6	7
8	9	10	11	12	13	14
15	16	17	18	19	20	21
22	23	24	25	26	27	28
29	30	31				

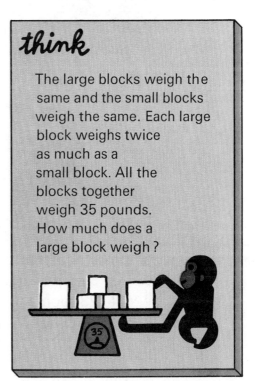

think

The large blocks weigh the same and the small blocks weigh the same. Each large block weighs twice as much as a small block. All the blocks together weigh 35 pounds. How much does a large block weigh?

Improving Multiplication Skills

1. Find the products.

A 37 ×4	B 56 ×2	C 95 ×3	D 43 ×4	E 82 ×5	F 96 ×4
G 54 ×3	H 38 ×6	I 69 ×5	J 57 ×2	K 70 ×3	L 58 ×2
M 49 ×5	N 68 ×4	O 65 ×4	P 85 ×4	Q 66 ×2	R 77 ×3
S 72 ×6	T 48 ×5	U 75 ×4	V 57 ×6	W 79 ×2	X 59 ×3

2. Find the area of this rectangle.

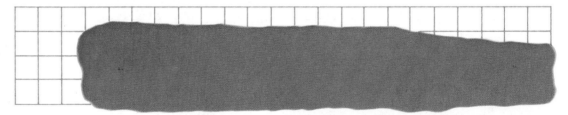

3. How many rooms in each?

A 7 stories high. 25 rooms on each floor.

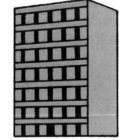

B 9 stories high. 23 rooms on each floor.

think

Jim had 1 minute to decide which of these allowances he wanted.

▶ $1.00 per week

or

▶ Each week he gets 1¢ the first day, 2¢ the second, 4¢ the third, and so on for 7 days.

Which would you take?

Give yourself 1 minute to decide. Then figure it out.

Solving Story Problems

A MOON TRIP

Peter thinks he would like to be an astronaut and go to the moon. He read some books to learn more about the moon. Peter wrote this paper to show his teacher.

Moon Facts

1. The United States Astronauts, Neil A. Armstrong and Edwin E. Aldrin, Jr., were the first men to land on the moon, July 20, 1969.
2. People weigh 6 times as much on earth as they do on the moon.
3. The moon goes around the earth once in about 28 days.
4. The moon is about 240,000 miles from the earth.
5. Scientists have found no air or water on the moon.

Peter L.

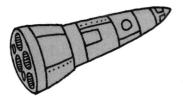

1. Peter figured he would weigh only 12 pounds on the moon. How much does Peter weigh on earth?

2. Peter's father would weigh 27 pounds on the moon. How much does he weigh on earth?

3. About how long does it take the moon to go around the earth 4 times?

4. Suppose you could fly straight to the moon and back. How far would you travel?

5. Peter drinks 4 cups of water each day.
   A How much water would he have to take for a 36-day moon trip?
   B How many quarts would this be?
   C How many gallons?

● *Let's explore larger products.*

Investigating the Ideas

Use the products on the three cards to help you give the three products below.

```
 3 0 0        2 0          6
 × 3         × 3         × 3
 9 0 0        6 0         1 8
```

A) 2 6
   × 3

B) 3 2 0
    × 3

C) 3 2 6
    × 3

 Can you make some cards like these that would help a classmate find this product?

2 4 3
× 4

Discussing the Ideas

1. Explain each step in the example below.

Step 1	Step 2	Step 3	Step 4
237 × 4 ――― 28	237 × 4 ――― 28 120	237 × 4 ――― 28 120 800	237 × 4 ――― 28 120 800 ――― 948
4 × 7 = 28	4 × 30 = 120	4 × 200 = 800	28 + 120 + 800 = 948

2. You know the first product. Find each of the other products.

   237
   × 4
   ―――
   948

   A) 2000
       × 4

   B) 2237
       × 4

   C) 3237
       × 4

248

Using the Ideas

Find the products.

1.	231 ×3	**2.**	213 ×4	**3.**	116 ×5	**4.**	326 ×2	**5.**	207 ×3
6.	128 ×4	**7.**	382 ×2	**8.**	144 ×3	**9.**	143 ×6	**10.**	162 ×5
11.	211 ×9	**12.**	264 ×3	**13.**	225 ×4	**14.**	243 ×6	**15.**	415 ×5
16.	1345 ×4	**17.**	2534 ×3	**18.**	1023 ×8	**19.**	1203 ×9	**20.**	1620 ×6

Solving Story Problems

1. Miss Wright took her pupils to visit the airport. While they were there, 3 jet planes took off for Europe. Each plane carried 125 people. How many people is this in all?

2. The large jet airliners have 4 engines. If each engine gives about 16,000 pounds of thrust (push) during takeoff, what is the total thrust then?

3. During most of the trip, each engine gives about 6000 pounds of thrust. What is the total thrust then?

4. Suppose a large jet plane flies at the rate of 575 miles each hour. How far would it fly in 4 hours?

5. A pilot told the children that some supersonic planes fly 3 times as fast as jets fly now. How fast is this? (Use the speed in exercise 4.)

More practice, page A-33, Set 44

• Can you estimate products like 4 × 49?

Investigating the Ideas

Sara is estimating the product 4 × 49. When we estimate an answer to a problem, we try to find a number that is "close" to the correct answer.

Work with one or more classmates. Choose one number from Set A and another from Set B. Each of you write down your estimate of the product.

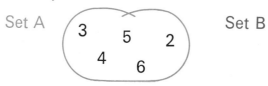 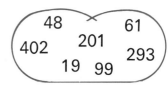

 Can you find the difference between your estimate and the actual product?

Discussing the Ideas

1. Explain how Sara found an estimate for 4 × 49.

2. Find the product in the red cloud. Explain why this is a "good estimate" of the product below.

   A (4 × 50)   B (3 × 40)   C (12 × 100)   D (6 × 200)
     4 × 51   3 × 39   12 × 99   6 × 204

3. Give an estimate for each product. Explain how you found your estimate.

   A 3 × 69   B 5 × 38   C 7 × 21   D 6 × 98   E 7 × 297

250

Using the Ideas

Estimate each product.

1. 3 × 82
2. 4 × 69
3. 8 × 99
4. 25 × 99
5. 6 × 61
6. 3 × 91
7. 5 × 199
8. 6 × 58
9. 2 × 197
10. 4 × 299
11. 5 × 302
12. 8 × 493

Which of the three numbers is a "good estimate"?

13. An airplane flew 502 miles each hour for 4 hours. About how many miles did it fly? A 200   B 2000   C 1000

14. An orchard has 8 rows of trees with 49 trees in each row. About how many trees are in this orchard?
    A 100   B 500   C 400

15. Each of 6 books has 312 pages. About how many pages are there in all?   A 2000   B 600   C 3000

16. If there are 365 days in each of 3 years, about how many days is this in all?   A 100   B 1000   C 10,000

17. There are 5280 feet in 1 mile. About how many feet are in 2 miles?   A 1000   B 100,000   C 10,000

★ 18. There are 52 weeks in a year. About how many weeks are in $2\frac{1}{2}$ years?
    A 100   B 200   C 125

★ 19. There are 60 minutes in an hour. About how many minutes are in $8\frac{1}{4}$ hours?
    A 450   B 500   C 550

★ 20. There are 1760 yards in a mile. About how many yards are in $2\frac{1}{2}$ miles?
    A 4500   B 3500   C 5500

Reviewing the Ideas

1. Solve the equations.
   A $4 \times 28 = (4 \times 20) + (4 \times n)$
   B $2 \times 54 = (2 \times 50) + (2 \times n)$
   C $3 \times 65 = (3 \times n) + (3 \times 5)$
   D $7 \times 18 = (7 \times n) + (7 \times 8)$
   E $6 \times 51 = (6 \times 50) + (6 \times n)$
   F $8 \times 46 = (8 \times n) + (8 \times 6)$

2. Find the products.
   A 100×18
   B 78×10
   C 6×100
   D 2×10
   E 10×136
   F 37×10
   G 100×24
   H 75×100
   I 20×10
   J 75×10
   K 10×23
   L 4×100

3. Find the products.
   A 3×20
   B 3×40
   C 5×50
   D 6×30
   E 2×80
   F 9×20
   G 7×20
   H 4×30
   I 5×60
   J 3×70
   K 2×90
   L 4×60

4. Find the products and sums.
   A $3 \times 20 = n$
   $3 \times 4 = n$
   $(3 \times 20) + (3 \times 4) = n$
   $3 \times 24 = n$

   B $4 \times 30 = n$
   $4 \times 6 = n$
   $(4 \times 30) + (4 \times 6) = n$
   $4 \times 36 = n$

5. Find the products.
   A 23×2
   B 14×2
   C 32×3
   D 12×4
   E 11×5
   F 43×2
   G 13×3
   H 22×4
   I 33×2
   J 24×2
   K 11×6
   L 12×3

6. Find the products.
   A 37×4
   B 46×3
   C 25×2
   D 39×5
   E 43×6
   F 62×4
   G 92×3
   H 49×4
   I 65×2
   J 238×3
   K 1292×4
   L 2158×5

★ **7.** Complete each sentence.
   A Since 4 × 27 = 108, we know that 5 × 27 = 108 + *n*.
   B Since 6 × 78 = 468, we know that 7 × 78 = 468 + *n*.
   C Since 6 × 78 = 468, we know that 5 × 78 = 468 − *n*.
   D Since 9 × 57 = 513, we know that 8 × 57 = 513 − *n*.
   E Since 7 × 83 = 581, we know that 8 × 83 = 581 + *n*.
   F Since 8 × 47 = 376, we know that 7 × 47 = 376 − *n*.
   G Since 4 × 348 = 1392, we know that 5 × 348 = 1392 + *n*.
   H Since 9 × 176 = 1584, we know that 8 × 176 = 1584 − *n*.
   I Since 27 × 38 = 1026, we know that 28 × 38 = 1026 + *n*.
   J Since 85 × 67 = 5695, we know that 84 × 67 = 5695 − *n*.

Short Stories

 13 rows of chairs.
6 chairs in each row.
How many chairs?

 3 classes of children.
34 children in each class.
How many children?

 427 pages in each book. 7 books. How many pages?

36 rooms on each floor.
9 floors. How many rooms?

 10 pencils in each bundle.
68 bundles.
How many pencils?

 7 dozen eggs. How many eggs?

 4 cups in a quart.
83 quarts.
How many cups?

12 inches in a foot.
7 feet.
How many inches?

 36 inches in a yard. How many inches in 10 yards?

100-yard dash. 3 feet per yard.
How many feet in the dash?

Keeping in Touch with Addition Subtraction Multiplication Division Fractions

1. Find the sums and differences.

   A 34   B 76   C 67   D 87   E 15   F 95
     +51   −24   +12   −32   +84   −13

   G 67   H 132   I 76   J 161   K 85   L 153
     +89   −57   +98   −82   +67   −95

2. Find the missing factors.
   A $6 \times n = 42$ D $6 \times n = 48$ G $3 \times n = 27$ J $9 \times n = 0$
   B $n \times 5 = 15$ E $n \times 4 = 36$ H $6 \times n = 36$ K $7 \times n = 49$
   C $8 \times n = 32$ F $n \times 5 = 35$ I $n \times 7 = 42$ L $n \times 3 = 9$

3. Find the quotients.
   A Since $8 \times 9 = 72$, we know that $72 \div 9 = n$.
   B Since $8 \times 9 = 72$, we know that $72 \div 8 = n$.
   C Since $6 \times 8 = 48$, we know that $48 \div 6 = n$.
   D Since $10 \times 27 = 270$, we know that $270 \div 27 = n$.

4. There are 12 children playing kickball.
   A One half of them are girls. How many of the children are girls?
   B One fourth of the children wear glasses. How many of the children wear glasses?

★ 5. Copy the problems and give the missing digits.

   A   59   B   39   C   83   D   23   E ▓9▓   F ▓▓▓
     −▓▓▓   +6▓   −3▓   ×▓   +3▓8   ×6
       28   ▓▓6   ▓6   46   860   252

6. Find the quotients.

A	12 ÷ 4	L	15 ÷ 3
B	20 ÷ 5	M	14 ÷ 7
C	12 ÷ 3	N	20 ÷ 4
D	18 ÷ 6	O	12 ÷ 6
E	10 ÷ 2	P	25 ÷ 5
F	15 ÷ 5	Q	24 ÷ 8
G	16 ÷ 4	R	21 ÷ 3
H	18 ÷ 3	S	28 ÷ 4
I	12 ÷ 2	T	36 ÷ 9
J	30 ÷ 5	U	30 ÷ 6
K	24 ÷ 6	V	35 ÷ 5

think

There are 16 third-graders in Pam's club at school. There are 5 third-grade teachers in the school. Explain why at least 4 of the third-grade children have the same teacher.

7. A group of 52 girls and 35 boys visited the bakery. They left at 9 o'clock in the morning and returned at 2 o'clock in the afternoon.
   A How many children went on the trip?
   B How many more girls went than boys?
   C How long did their trip last?
   D Only 23 girls rode the bus. How many did not ride the bus?
   E The boys visited one room in groups of 7. How many groups of 7 boys were there?
   F Each of the girls got 7 souvenir pencils. How many pencils did they get in all?
   G Each of the boys got 7 souvenir pencils. How many pencils did they get in all?
   H How many more pencils did the girls get than the boys?
   I How many pencils did the children receive in all?

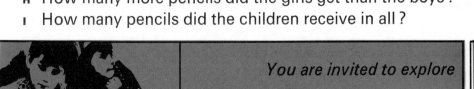

You are invited to explore

ACTIVITY CARD 12
Page 315

11 Geometry and Graphing

● *Can number pairs show location?*

Investigating the Ideas

Study the graphs.

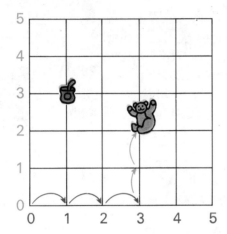
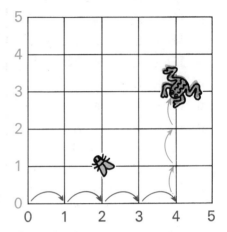

The bear is
"3 over and 2 up."
Its coordinates are (3,2).

The frog is
"4 over and 3 up."
Its coordinates are (4,3).

 Can you answer these questions about the graphs?

1. A Where is the honey?
   B What are its coordinates?

2. A Where is the fly?
   B What are its coordinates?

Discussing the Ideas

1. Does "2 over and 3 up" give the same location as (3,2)?

2. Explain how you would find the location for each of these coordinates.
   A (2,5)   B (5,4)   C (3,1)

Using the Ideas

1. Give the missing numbers.
   A The butterfly is 2 over and ? up.
   B The grasshopper is ? over and 4 up.
   C The bee is ? over and ? up.
   D The spider is ? over and ? up.
   E The ladybug is ? over and ? up.

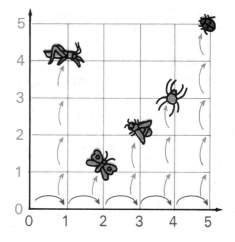

2. Give the coordinates for each insect above.

3. Give the missing numbers. Then give the coordinates.
   A The ball is 4 over and ▒ up. The coordinates for the ball are __?__.
   B The bird is ▒ over and 7 up. The coordinates for the bird are __?__.
   C The apple is 7 over and ▒ up. The coordinates for the apple are __?__.
   D The car is 6 over and ▒ up. The coordinates for the car are __?__.
   E The block is ▒ over and 5 up. The coordinates for the block are __?__.

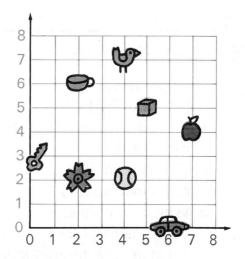

   F The key is 0 over and ▒ up. The coordinates for the key are __?__.
   G What are the coordinates for the cup?
   H Give the coordinates for the flower.

• Can you find coordinates of points?

Investigating the Ideas

The letters on the graph form a three-word secret message.

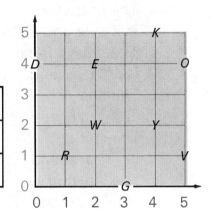

First Word:	(5,1), (2,4), (1,1), (4,2)
Second Word:	(3,0), (5,4), (5,4), (0,4)
Third Word:	(2,2), (5,4), (1,1), (4,5)

 Can you find the secret message?

Discussing the Ideas

1. The coordinates of the point P are (1,4).
   A What are the coordinates of point Q?
   B Which point has coordinates (2,1)?
   C Which point has coordinates (1,2)?
   D What are the coordinates of point W?
   E Which point has coordinates (0,3), M or N?

2. A What can be said about the locations of points N, R, S, and M?
   B What can be said about the coordinates of these points?

Using the Ideas

1. Answer the questions.

   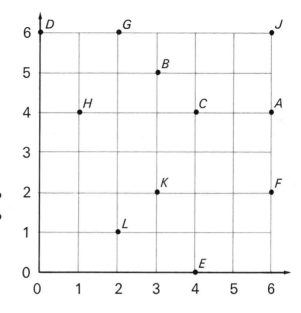

   A What letter is 6 over and 4 up?
   B What letter is 3 over and 5 up?
   C What letter has coordinates (6,4)?
   D What letter has coordinates (3,5)?
   E What letter is at (4,4)?
   F What letter is at (1,4)?
   G What letter has coordinates (4,0)?
   H Coordinates (0,6) locate what letter?
   I What do you find at (2,6)?
   J Coordinates (6,2) locate what letter?

2. What are the coordinates of J, K, and L in exercise 1?

think

Can you write a secret message on a grid, as in the Investigation? Give the coordinates for your secret message to a classmate. See if he can find your message.

● How do you graph a point?

Discussing the Ideas

1. Jill started at 0, counted over 4 and then up 2 to find the location of the point. Could you show her an easier way?

2. Here are some other points Jill graphed. Explain how she might have counted to decide where to mark each point. What are the coordinates of each point?

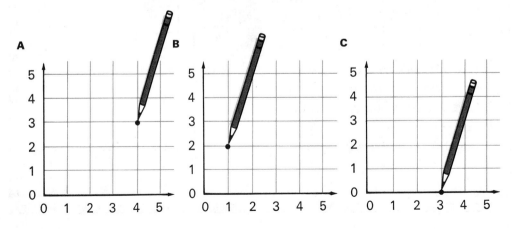

Using the Ideas

1. Label your graph paper like this. Then graph each of these points and write the letter beside it.

   A(7,3)   E(8,1)
   B(3,7)   F(0,3)
   C(5,5)   G(2,1)
   D(1,6)   H(6,0)

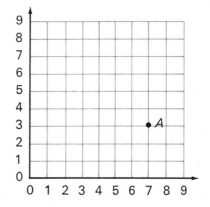

2. **a** Give the coordinates of points A, B, and C.
   b What figure is formed?
   c On your paper, graph three other points to form a triangle. Give the coordinates of your points.
   d Give the coordinates of three points that **cannot** be connected to form a triangle.

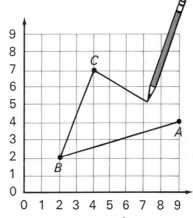

3. **a** Give the coordinates of points A and B.
   b Give the coordinates of two more points needed to form a square.
   c On your paper, graph four other points to form a square. Give the coordinates of your points.
   ★ **d** Give the coordinates of the points needed to form another figure. Graph and name the figure.

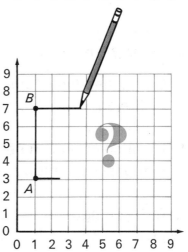

261

● How can you make point pictures?

Investigating the Ideas

This teepee was made by graphing and connecting points with these coordinates.
(1,1) → (3,6) → (5,1) → (4,1) → (3,2) → (2,1) → (1,1)
The arrows show the order of the points.

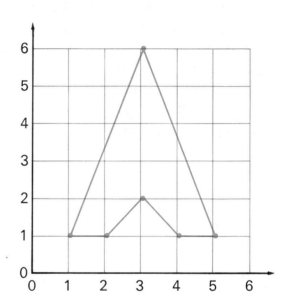

Can you make a picture by graphing and connecting these points in order? Try it.
(0,2) → (3,6) → (5,5) → (6,2) → (4,3) → (5,1) → (3,3) → (2,0) → (0,2)

Discussing the Ideas

1. Could the points for the teepee be graphed and connected in a different order? Explain.

2. What point do you need to complete the picture you made in the Investigation? What are its coordinates?

3. Can you figure out the coordinates of points that will give you one of these familiar geometric figures?
   A square   B parallelogram   C rhombus

Using the Ideas

1. If these remaining 6 points are graphed and connected, the airplane will be finished.
   $(9,0) \rightarrow (6,4) \rightarrow (2,1) \rightarrow (3,0) \rightarrow (1,0) \rightarrow (0,1)$
   Copy and complete the picture on your graph paper.

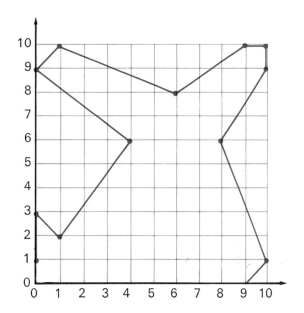

2. The first three points have been graphed and connected. Copy and complete the picture on your graph paper.
   $(0,3) \rightarrow (2,3) \rightarrow (7,9) \rightarrow (10,10) \rightarrow (10,9) \rightarrow (9,8) \rightarrow (10,8) \rightarrow (3,2) \rightarrow (3,0) \rightarrow (2,2) \rightarrow (0,3)$

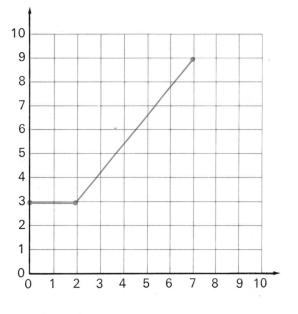

★ 3. Draw a picture by connecting points on your graph paper. Give the coordinates, in order, to a classmate and see if he can draw the picture.

● Let's explore symmetry on graph paper.

Investigating the Ideas

The paper has been folded so that only half of a symmetric figure can be seen.

Copy this part of the figure on your graph paper.

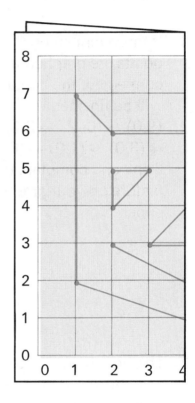

 Can you show the other half of the picture on your graph?

Discussing the Ideas

1. A What are the coordinates of the point at the tip of the cat's ear shown above?
   B What are the coordinates of the point at the tip of the cat's other ear?

2. Pick other points and give their coordinates. Then give the coordinates of the matching points in the other half of the picture.

3. Where is the line of symmetry of the figure?

264

Using the Ideas

1. Use your graph paper to show what each symmetric picture will look like when the paper is unfolded.

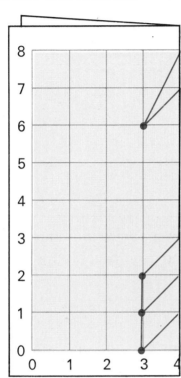
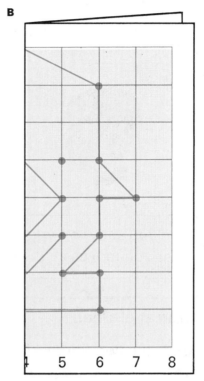

2. A Give the coordinates of each of the four points marked in 1 A.
   B Give the coordinates of four matching points in the other half of the picture for 1 A.

★ 3. Use graph paper to make a symmetric picture of your own. Then fold it in half and see if a classmate can draw the complete picture.

You think about 10 hundreds.
100 tens will do.
That's enough to find my name.
You need no other clue.

WHO AM I?

● *How can figures be moved on a graph?*

Investigating the Ideas

Jon

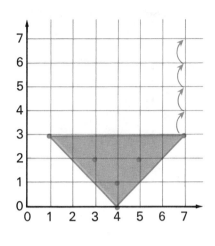

 Can you show on your graph paper where the figure will be after Jon moves all the points and connects them?

Discussing the Ideas

1. Look at the two figures. Did Rachel follow the directions?

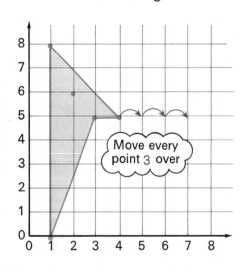
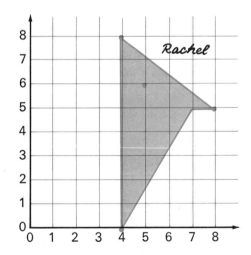

2. Can you show on your graph paper how Rachel should have drawn the figure?

266

Using the Ideas

Use the moves given and show the final position of each picture on your graph paper.

1.

Move every point 2 up ↰.

2.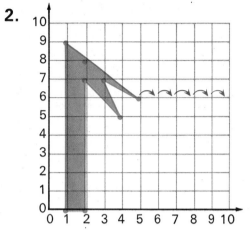

Move every point 5 over ↷.

3.

Move every point 3 down ↓.

★ 4.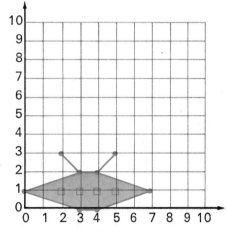

Move every point 3 over and 7 up.

★ 5. Make a figure of your own on graph paper and then move it 3 over **and** 2 up.

● *Can input-output pairs be shown on a graph?*

Investigating the Ideas

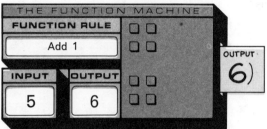

This machine puts out an **input-output** card each time it operates.

The point for the first card is shown on the graph.

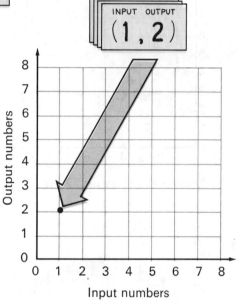

 Can you write the number pairs for 5 more input-output cards and show the point for each pair on a graph?

Discussing the Ideas

1. If the input number were 0, where would you mark the point on your graph?

2. Did all the points you graphed lie in a straight line?

3. Do you think the points would lie in a straight line no matter what rule was used?

Using the Ideas

1. Use the function table to write the coordinates for points *B* through *G*. Then graph the point for each of the coordinates on your graph paper. Point *A* is graphed correctly.

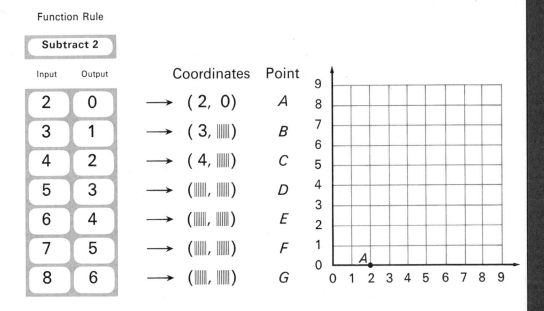

2. Think about input-output cards and graph the points for each table.

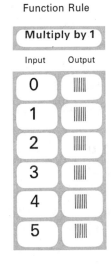

• How do you make bar graphs?

Investigating the Ideas

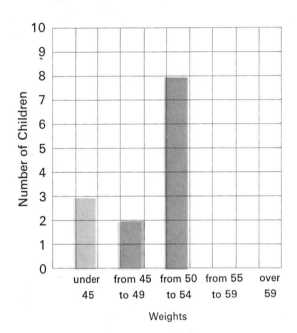

Information			
Amy	41	Kent	53
Beth	43	Lisa	54
Carl	44	Mike	54
Don	46	Nan	55
Eric	49	Orin	57
Fran	50	Paul	57
Gail	51	Ray	58
Hal	51	Sara	59
Ivan	52	Ted	63
Jill	53	Val	66

Lisa colored each bar on her graph paper to show how many children had the weight given below the bar.

 Can you tell how to make the last two bars to finish Lisa's graph?

Discussing the Ideas

1. Do you think the bar graph is a good way to show the information about weights? Why?

2. How can you tell from the bar graph how many students weigh from 50 to 54 pounds?

3. Collect information about the weights of children in your class. How would the bar graph for your class look?

Using the Ideas

1. Draw and color bars on graph paper to show the number of people in the family of each student.

Name	Number of people in family
Ann	4
Bob	3
Cal	5
Don	7
Eve	10
Flo	2

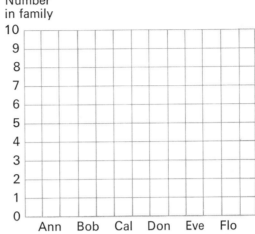

2. Make a bar graph like the one in exercise 1 for eight students in your class.

3. Make a bar graph using the idea suggested by the picture.

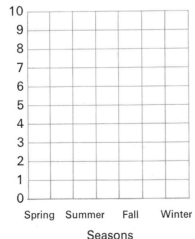

think

I'm not so very large,
Just over 33.
Add some fives together,
My name you soon will see.

WHO AM I?

ADD FIVES

ADD SIXES

I'm a little over 16
And less than 22.
If you'll just start adding sixes,
My name will come to you.

WHO AM I?

● *How can we use negative numbers in graphing?*

Investigating the Ideas

Alice made this graph to show the temperature from 1:00 A.M. to 8:00 A.M. on a very cold day. She used **negative** numbers to show temperatures **below** zero and positive numbers to show temperatures **above** zero.

For negative numbers, we read: Negative one, negative two, negative three, . . .

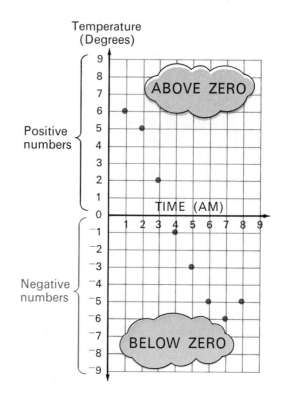

 Can you record the temperatures shown for each of the hours on the graph?

Discussing the Ideas

1. A What was the temperature at 1:00 A.M.?
   B What was the temperature at 2:00 A.M.?
   C When was the temperature 2 degrees?
   D What would the coordinates (4, −1) tell you?
   E At what time was the temperature 3 degrees below zero?
   F What is the coldest temperature shown on the graph?
   G Look at the graph and **guess** what the temperature would be at 9:00 A.M.

2. Can you think of some other ways that negative numbers might be used?

Using the Ideas

1. Ted made this graph to show the changes in pulse rate of an astronaut around blast-off time. He used **negative** numbers for the seconds **before** blast-off and positive numbers for the seconds **after** blast-off.

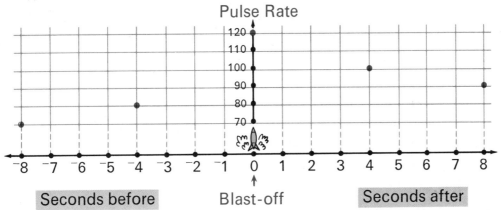

Seconds before   Blast-off   Seconds after

A. What was the pulse rate 8 seconds before blast-off? What are the coordinates of that point?

B. Give the coordinates for the pulse rate 4 seconds before blast-off.

C. What was the pulse rate at blast-off? Give the coordinates of that point.

D. Give the coordinates for each of the other points on Ted's graph.

2. Give the missing numbers in the table. The graph may help you.

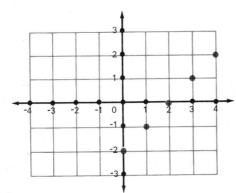

Function Rule

Subtract 2

Input	Output
4	2
3	1
2	0
1	
0	

Reviewing the Ideas

1. **A** What letter is 1 over and 2 up?
   B The point lettered G has what coordinates?
   C What are the coordinates for point A?
   D What is the letter for the point with coordinates (2,5)?
   E Are the coordinates of point E (3,0) or (0,3)?

2. Graph the points in the order listed and connect them to form a picture.
   A (3,0) → (3,1) → (2,1) → (1,3)
   → (2,2) → (4,2) → (6,4) → (7,3)
   → (6,3) → (5,1) → (4,1) → (4,0)
   B (2,0) → (2,1) → (1,3) → (2,5)
   → (2,6) → (3,5) → (4,5) → (5,6)
   → (5,5) → (6,3) → (5,1) → (5,0)

3. On graph paper, show how the figure would look for each move.
   A Move every point 2 over.
   B Move every point 3 up.

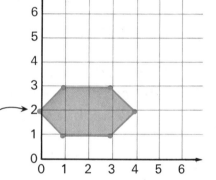

4. Complete the function tables and graph the input-output pairs.

A Function Rule: Subtract 2

Input	Output
8	6
7	5
6	
5	
4	

B Function Rule: Multiply by 2

Input	Output
0	0
1	2
2	
3	
4	

C Function Rule: Add 0

Input	Output
1	
2	
3	
4	
5	

Keeping in Touch with   Addition   Multiplication
                                         Subtraction   Division

1. Write a multiplication equation for each exercise.

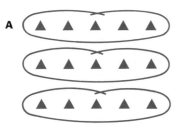    C $6 + 6 + 6 + 6 + 6$

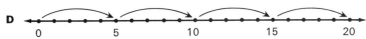

2. Write 2 division equations for each exercise.

    A $4 \times 5 = 20$ B $7 \times 9 = 63$ C $17 \times 28 = 476$

3. Find the products.

	A	B	C	D	E	F	G	H
	7	8	3	9	5	4	6	7
	×6	×5	×8	×7	×7	×6	×8	×7

	I	J	K	L	M	N	O	P
	9	7	4	9	7	8	8	5
	×6	×4	×9	×8	×8	×4	×8	×6

4. Find the sums and differences.

	A	B	C	D	E	F
	27	64	24	64	64	59
	+44	−31	+65	−34	−38	+63

	G	H	I	J	K	L
	47	93	127	141	65	48
	+56	−26	−54	−95	−19	+16

   *You are invited to explore*   **ACTIVITY CARD 13** Page 315

12 Dividing

● *What are some ways to think about division?*

Investigating the Ideas

You can use sets, the number line, or subtraction to help you find quotients.

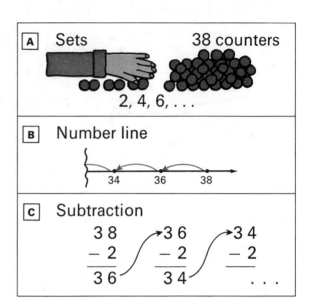

 Can you use one of these methods to help you find how many twos are in 38?

Discussing the Ideas

1. How would you show 24 ÷ 6 using method A?

2. Explain how you would show 12 ÷ 4 on a number line.

3. Give each difference for finding 36 ÷ 9 by subtraction.

4. What division problem can you solve if you know the missing factor in the equation?

$$? \times 5 = 30$$

276

Using the Ideas

1. Write a division equation for each exercise.

18 stars

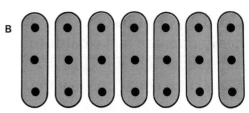

21 dots

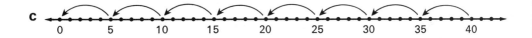

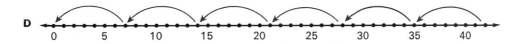

E 30 − 6 = 24
24 − 6 = 18
18 − 6 = 12
12 − 6 = 6
6 − 6 = 0

F 24 − 8 = 16
16 − 8 = 8
8 − 8 = 0

G 20 − 5 = 15
15 − 5 = 10
10 − 5 = 5
5 − 5 = 0

H 3 × 4 = 12   I 9 × 8 = 72   J 5 × 7 = 35

2. Find the products.
   A 3 × 8   E 4 × 8   I 2 × 9   M 1 × 7   Q 1 × 9
   B 6 × 7   F 0 × 8   J 3 × 7   N 6 × 4   R 8 × 3
   C 5 × 5   G 5 × 4   K 5 × 6   O 5 × 9   S 9 × 4
   D 3 × 9   H 6 × 6   L 7 × 3   P 4 × 7   T 6 × 8

3. Find the quotients.
   A 18 ÷ 9   E 42 ÷ 7   I 24 ÷ 8   M 24 ÷ 6   Q 36 ÷ 4
   B 7 ÷ 7   F 0 ÷ 8   J 24 ÷ 4   N 21 ÷ 7   R 45 ÷ 9
   C 20 ÷ 4   G 21 ÷ 3   K 36 ÷ 6   O 30 ÷ 6   S 18 ÷ 2
   D 25 ÷ 5   H 48 ÷ 6   L 27 ÷ 9   P 28 ÷ 7   T 32 ÷ 8

● Can products be used to find quotients?

Investigating the Ideas

Use 3-by-5 cards or pieces of heavy paper to make cards for a Product-Quotient Quiz. Make up 8 equations like the ones shown, but with different numbers.

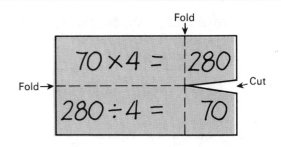

As one factor, use a number ending in 0, such as 70, 80, 500, or 700. For the other factor, choose from the digits 2, 3, 4, 5, 6, 7, 8, and 9. Cut and fold your 8 cards as shown.

 Can you follow the directions below to quiz one of your classmates?

| Question 1 | Check. | Question 2 | Check. |

Discussing the Ideas

1. How can you use the multiplication equation to help you solve the division equation?

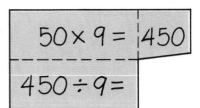

2. How would you complete this card?

278

Using the Ideas

1. Find the products.
   - A 5×90
   - B 80×8
   - C 7×80
   - D 70×7
   - E 70×6
   - F 6×50
   - G 5×50
   - H 3×80
   - I 9×70
   - J 60×6
   - K 8×50
   - L 30×9
   - M 50×7
   - N 6×80
   - O 30×2
   - P 60×3
   - Q 8×80
   - R 4×80
   - S 6×40
   - T 5×60

2. Find the quotients.
   - A $450 \div 5$
   - B $640 \div 8$
   - C $560 \div 7$
   - D $300 \div 6$
   - E $250 \div 5$
   - F $360 \div 6$
   - G $270 \div 9$
   - H $180 \div 3$
   - I $490 \div 7$
   - J $630 \div 9$
   - K $480 \div 6$
   - L $320 \div 4$

3. Find the products.
   - A 9×400
   - B 300×8
   - C 5×700
   - D 3×900
   - E 800×5
   - F 500×5
   - G 8×800
   - H 700×8
   - I 7×700
   - J 300×3
   - K 3×500
   - L 600×8
   - M 400×8
   - N 4×500
   - O 6×300
   - P 6×600

think

Now here's a simple rule
That should prove quite
   a friend.
When you multiply by me,
Put two zeros on the end.
WHO AM I?

4. Find the missing factors.
   - A $5 \times n = 350$
   - B $n \times 7 = 490$
   - C $4 \times n = 200$
   - D $n \times 6 = 420$
   - E $9 \times n = 630$
   - F $n \times 8 = 2400$
   - G $7 \times n = 5600$
   - H $n \times 8 = 3200$
   - I $5 \times n = 4500$
   - J $n \times 2 = 600$
   - K $9 \times n = 450$
   - L $n \times 8 = 6400$
   - M $3 \times n = 2100$
   - N $n \times 5 = 300$
   - O $6 \times n = 2400$

5. Find the quotients.
   - A $400 \div 8 = n$
   - B $180 \div 6 = n$
   - C $450 \div 5 = n$
   - D $560 \div 7 = n$
   - E $630 \div 7 = n$
   - F $600 \div 3 = n$
   - G $2400 \div 4 = n$
   - H $3500 \div 5 = n$
   - I $2000 \div 4 = n$
   - J $6300 \div 9 = n$
   - K $900 \div 3 = n$
   - L $2500 \div 5 = n$
   - M $3600 \div 6 = n$
   - N $3000 \div 5 = n$
   - O $4000 \div 5 = n$

More practice, page A-34, Set 45

What's the input, output, or rule?

Study the picture. Then give the number or function rule for each gray space in exercises 1 through 6.

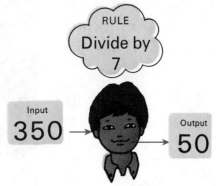
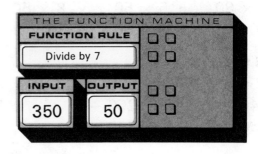

1. Function Rule: Multiply by 10

	Input	Output
	15	150
A	24	
B	37	
C	48	
D		720

2. Function Rule: Multiply by 100

	Input	Output
	8	800
	23	2300
A	65	
B	83	
C		2800

3. Function Rule

	Input	Output
A		
	9	360
	4	160
	6	240
B	5	
C		280

4. Function Rule: Divide by 4

	Input	Output
	80	20
A	280	
B	240	
C	120	
D	200	

5. Function Rule: Divide by 3

	Input	Output
	60	20
	180	60
A	240	
B	120	
C		30

★ 6. Function Rule

	Input	Output
A		
	4	16
	5	25
	10	100
B	8	
C		49

280

Let's use a special function machine.

This special function machine uses the output number as a new input number and keeps operating. A counter tells how many times the rule is used.

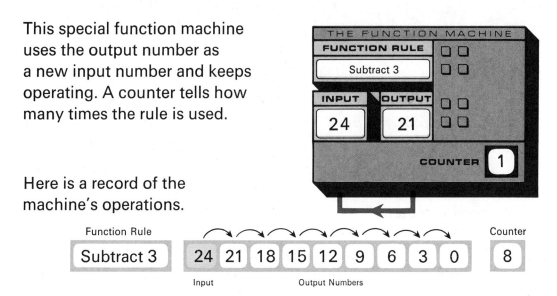

Here is a record of the machine's operations.

Function Rule				Input			Output Numbers					Counter
Subtract 3				24	21	18	15	12	9	6	3	0	8

Here are more records. Give what you think should go in each gray space, then write a division equation for the exercise.

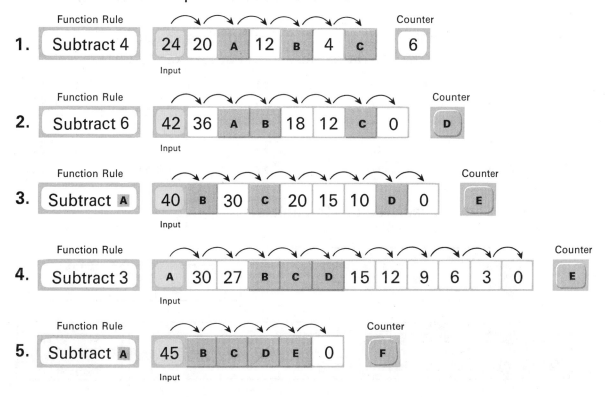

Keeping in Touch with Subtraction   Measurement
                          Multiplication Story problems

1. Give the mark (>, <, =) that should go in each ◉.
   - A 70 × 3 ◉ 3 × 70
   - B 3 × 40 ◉ 6 × 20
   - C 70 × 5 ◉ 9 × 40
   - D 99 × 4 ◉ 4 × 100
   - E 5 × 50 ◉ 6 × 40
   - F 98 × 7 ◉ 700
   - G 8 × 60 ◉ 70 × 7
   - H 700 × 8 ◉ 600 × 9
   - I 30 × 7 ◉ 70 × 3
   - J 800 × 9 ◉ 900 × 8
   - K 48 × 8 ◉ 8 × 53
   - L 34 × 5 ◉ 35 × 4

2. Find the missing factors.
   - A $n \times 6 = 240$
   - B $n \times 4 = 240$
   - C $4 \times n = 320$
   - D $8 \times n = 480$
   - E $n \times 9 = 360$
   - F $n \times 9 = 270$
   - G $n \times 7 = 350$
   - H $7 \times n = 420$
   - I $5 \times n = 400$
   - J $5 \times n = 450$
   - K $n \times 7 = 210$
   - L $n \times 6 = 300$
   - M $3 \times n = 270$
   - N $3 \times n = 120$
   - O $n \times 8 = 560$
   - P $n \times 6 = 480$
   - Q $n \times 6 = 180$
   - R $3 \times n = 150$

3. There are no whole number answers for 3 of these exercises. List these. Then find the differences in the other exercises.
   - A 65 − 23
   - B 49 − 26
   - C 31 − 42
   - D 548 − 234
   - E 156 − 133
   - F 657 − 756
   - G 329 − 216
   - H 254 − 237
   - I 472 − 481
   - J 355 − 167

4. Find the area of each region. The unit is ▫.

   A   B

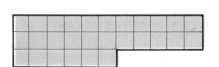
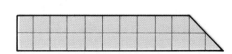

You are invited to explore **ACTIVITY CARDS 14, 15** Page 316

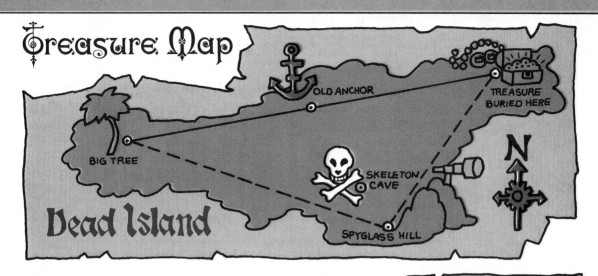

5. John found this map in an old trunk in his attic. One piece of paper in the trunk where the map was found looked like this.

   Scale A: 1 inch
   Each inch on the map means 1 mile on the island.

   Use scale A to answer these questions.
   A   How many miles is it from Big Tree to Spyglass Hill?
   B   How many miles is it from Spyglass Hill to the treasure?

   There was another scrap of paper in the trunk. Use scale B to answer the rest of the questions.

   Scale B: 1 inch
   Each inch on the map means 2 miles on the island.

   C   How many miles is it from Big Tree to Spyglass Hill?
   D   How many miles is it from Spyglass Hill to the treasure?
   E   How many miles would you walk if you went from Big Tree to Spyglass Hill and then to the treasure?
   F   How many miles is it from Big Tree straight to the treasure?
   G   How many miles is it from Old Anchor straight to Skeleton Cave?
   H   How far is it from Skeleton Cave to Spyglass Hill?
   I   Which is farther from Big Tree, Spyglass Hill or the treasure? How much farther?

● Can subtracting help you find the quotients?

Discussing the Ideas

1.  I'll subtract 1 four at a time to find out.

   $24 \div 4?$
   How many fours in 24?

   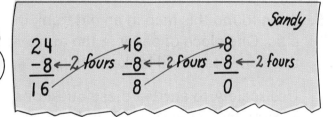

   A How many fours did Fred find in 24?
   B Solve: $24 \div 4 = n$

2.  I'll subtract 2 fours at a time to find how many fours are in 24.

   A How many fours did Sandy find in 24?
   B Whose method is shorter?

3.  I can do the problem by subtracting 3 fours at a time.

   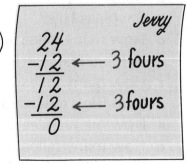

   A How many fours did Jerry find in 24?
   B Explain two ways Jerry's method is shorter than the others.

Using the Ideas

1. Find the differences. Then solve the division equation.

   A 25   20   15   10   5
      −5   −5   −5   −5   −5
      ——   ——   ——   ——   ——

      $25 \div 5 = n$

   B 18   15   12   9   6   3
      −3   −3   −3   −3   −3   −3
      ——   ——   ——   ——   ——   ——

      $18 \div 3 = n$

2. Find the differences. Then solve the division equation.

   A 21   B 16   C 27
      −7   −8   −9
      ——   ——   ——

      −7   −8   −9
      ——   ——   ——

      −7 $16 \div 8 = n$ −9
      ——   ——

   $21 \div 7 = n$ $27 \div 9 = n$

3. Find the differences. Then solve the division equation.

   A 48   B 35   C 30
      −16 ← 2 eights   −14 ← 2 sevens   −18 ← 3 sixes
      ———   ———   ———

      −16 ← 2 eights   −14 ← 2 sevens   −12 ← 2 sixes
      ———   ———   ———

      −16 ← 2 eights   −7 ← 1 seven $30 \div 6 = n$
      ———   ———

   $48 \div 8 = n$ $35 \div 7 = n$

• Can larger quotients be found by subtraction?

Investigating the Ideas

Find how many threes there are in 48 by starting with 48 and subtracting as many threes as you like each time.

```
How many threes
   in 48?
     48
    - ?
    ___

    - ?
```

 Can you find how many twos there are in 136?

Discussing the Ideas

A How many twos in 34? $34 \div 2$	B How many threes in 42? $42 \div 3$	C How many fours in 60? $60 \div 4$
34 −20 ← 10 twos ――― 14 −14 ← 7 twos ――― 0 $34 \div 2 = n$	42 −30 ← 10 threes ――― 12 −12 ← 4 threes ――― 0 $42 \div 3 = n$	60 −40 ← 10 fours ――― 20 −20 ← 5 fours ――― 0 $60 \div 4 = n$

1. What is the first step in each of the examples above? What is the second step?

2. How can you use these two steps to find each quotient?

Using the Ideas

1. Use the subtractions to help you find each quotient.

A. $36 \div 3 = n$

```
  36
- 30  ← 10 threes
  ───
   6
-  6  ← 2 threes
  ───
   0
```

B. $52 \div 4 = n$

```
  52
- 40  ← 10 fours
  ───
  12
- 12  ← 3 fours
  ───
   0
```

C. $84 \div 6 = n$

```
  84
- 60  ← 10 sixes
  ───
  24
- 24  ← 4 sixes
  ───
   0
```

2. Find the differences. Then find each quotient.

A. $46 \div 2 = n$

```
  46
- 20  ← 10 twos

- 20  ← 10 twos

-  6  ← 3 twos
```

B. $75 \div 3 = n$

```
  75
- 30  ← 10 threes

- 30  ← 10 threes

- 15  ← 5 threes
```

C. $126 \div 6 = n$

```
  126
-  60  ← 10 sixes

-  60  ← 10 sixes

-   6  ← 1 six
```

3. Find the quotients.

A. $92 \div 4 = n$
B. $93 \div 3 = n$
C. $148 \div 4 = n$
D. $78 \div 2 = n$
E. $115 \div 5 = n$
F. $177 \div 3 = n$
G. $96 \div 8 = n$
H. $185 \div 5 = n$
I. $162 \div 6 = n$
J. $224 \div 7 = n$

think

1. Pick a number.
2. Add 4.
3. Multiply by 2.
4. Subtract 6.
5. Divide by 2.
6. Subtract the number you started with.

Do you think you will always end with 1? Try this several times.

```
  ?
 +4
 ×2
 -6
 ÷2
 -?
 ───
```

• *Let's explore shorter ways for finding quotients.*

Discussing the Ideas

1. Copy each problem on paper or on the chalkboard. Subtract as many threes each time as shown by the number in the ring.

   A 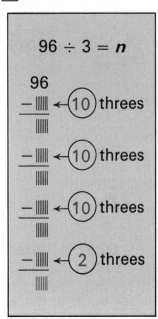 B 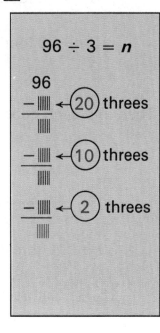 C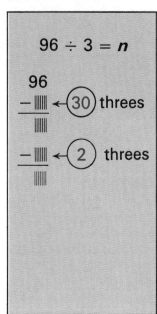

   Did you get 0 in the red screen each time?

2. A What is the quotient for each part above?
   B Did you get the same quotient for each part?
   C Which way was easiest?
   D Which way was shortest?

3. If you start with 48, can you subtract as many as
   A 10 twos? B 20 twos? C 30 twos?

4. A How many twos in all can you subtract from 48?
   B What is the quotient 48 ÷ 2?

Using the Ideas

1. Follow the directions. Give any missing numbers. Then find the quotients.

A
Start with 48.
⬇
Subtract (10) threes.
⬇
Subtract (6) threes to end with 0.

48 ÷ 3 = ?

B
Start with 96.
⬇
Subtract (20) fours.
⬇
Subtract (?) fours to end with 0.

96 ÷ 4 = ?

C
Start with 138.
⬇
Subtract (?) sixes.
⬇
Subtract (3) sixes to end with 0.

138 ÷ 6 = ?

2. Find the product. This will help you decide how many sevens you can subtract at first when finding the quotient. Write each quotient.

   A $20 \times 7 = n$
      $154 \div 7 = n$

   B $50 \times 7 = n$
      $371 \div 7 = n$

   C $80 \times 7 = n$
      $567 \div 7 = n$

3. Find these quotients.
   A 46 ÷ 2
   B 115 ÷ 5
   C 155 ÷ 5
   D 132 ÷ 4
   E 138 ÷ 3
   F 126 ÷ 6
   G 224 ÷ 7
   H 336 ÷ 8

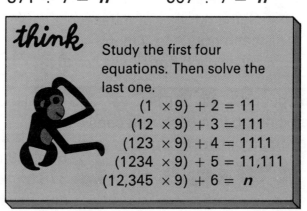

think

Study the first four equations. Then solve the last one.
$(1 \times 9) + 2 = 11$
$(12 \times 9) + 3 = 111$
$(123 \times 9) + 4 = 1111$
$(1234 \times 9) + 5 = 11{,}111$
$(12{,}345 \times 9) + 6 = n$

More practice, page A-35, Set 46

Solving Story Problems

1 300 baseball cards.
5 on each page.
How many pages?

2 320 airplane cards.
4 cards on each page.
How many pages?

3 96 marching-band players.
8 rows. Same number in each row. How many in each row?

4 To the zoo in buses.
210 children. 7 buses.
Same number in each bus.
How many children in each bus?

5 108 boys.
9 boys on each team.
How many teams?

6 Square dancing. 112 girls.
8 girls make a "square."
How many squares?

7 192 trading stamps. 8 pages.
Same number on each
page. How many stamps
on each page?

8 240 doll-picture cards.
6 on each page.
How many pages?

9 450 miles in 9 hours.
Same number of miles
each hour. How many
miles each hour?

10 216 pieces of candy.
6 pieces in each bag.
How many bags?

11 168 pennies.
Same number in each of 8 cans.
How many in each can?

12 Light bulb.
Blinks 9 times each minute.
378 blinks. How many
minutes have passed?

At the Scout Camp

1. 8 Girl Scouts slept in each cabin. How many cabins were used by the 56 girls from Pine City?

2. 54 Boy Scouts went to Camp Eagle in 6 station wagons. There were the same number of scouts in each station wagon. How many went in each station wagon?

3. There were 270 Girl Scouts at Camp Sunrise. There were 9 troops of the same size. How many were in each troop?

4. One week 208 scouts came to Camp Eagle. They lived in tents. 4 scouts slept in each tent. How many tents were used?

5. 200 Girl Scouts ate meals in a large cabin. 8 girls sat at each table. How many tables were there?

6. 72 scouts planned to take a boating trip from Camp Eagle to Camp Sunrise. If each boat could hold 6 scouts, how many boats were needed?

7. On July 4, a scoutmaster at Camp Eagle bought a bottle of soda for each scout. There were 540 Boy and Girl Scouts at the campfire party. How many cartons of 6 bottles did he buy?

8. During the summer 288 scouts visited an Indian museum. The guide took the scouts through the museum in groups of 9. How many trips did he make in all?

More practice, page A-35, Set 47

Keeping in Touch with — Addition, Subtraction, Multiplication, Fractions, Story problems

1. Write a fraction that shows the part of each region that is shaded.

   A 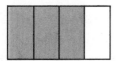   B 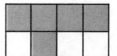   C

2. Find the products.

   A 34   B 26   C 42   D 37   E 73
      ×2   ×3   ×4   ×5   ×6

   F 68   G 59   H 74   I 83   J 92
      ×5   ×6   ×7   ×8   ×9

3. Answer "more than 100" or "less than 100" for each product.

   A 3 × 33   D 4 × 33   G 6 × 20   J 11 × 11
   B 3 × 34   E 5 × 21   H 10 × 9   K 9 × 11
   C 4 × 22   F 5 × 19   I 10 × 11   L 9 × 12

4. Find the sums.

   A 15   B 62   C 74
      32   24   37
      40   46   53
      ──   ──   ──

★ 5. Find the missing digits.

   A ▓5   B ▓8
     +4▓   +3▓
     ───   ───
      81   112

   C 6▓   D ▓4
     −▓8   −1▓
     ───   ───
      39   38

   E ▓6   F 3▓
      ×2   ×4
      ──   ──
      92   148

think

There were two stacks of checkers on the table. 6 checkers were removed from one stack and placed on the other. Then the stacks had the same number of checkers. Before moving the 6 checkers, how many more did the taller stack have than the shorter one?

6. Give the missing numbers.
   Examples:
   $3.47 means 3 dollars and 47 cents.
   $3.47 means 347 cents.
   A $5.39 means ▒ dollars and ▒ cents.
   B $3.86 means ▒ dollars and ▒ cents.
   C $4.23 means 3 dollars and ▒ cents.
   D $7.33 means ▒ cents.

1 dollar is worth 100 pennies.

7. Find the total amounts.
   Example: $1.34 and $3.23 is $4.57.
   A $3.32 and $2.45
   B $12.57 and $8.32
   C $3.64 and $4.58
   D $8.64 and $1.36
   E $ 7.32
       12.68
       ─────

8. Find the difference in the amounts.

Example:	A	B	C	D
$4.65	$5.86	$9.47	$8.52	$4.38
3.13	2.42	2.13	3.36	2.53
$1.52				

9. A Susan bought a record and a book.
      The clerk wrote the costs down like this.
      Find the total amount.
   B Susan gave the clerk 6 dollars.
      How much change did she get?
   C Susan had $8.67 when she left home. She
      spent $6.34. How much did she have left?

Record.... 3.49
Book..... 2.35
Total _____

10. Ken bought a baseball and a bat.
    Find the total cost.

$ 3.36 $ 1.89

You are invited to explore

ACTIVITY CARD 16
Page 317

● *Let's explore ways to write division exercises.*

Discussing the Ideas

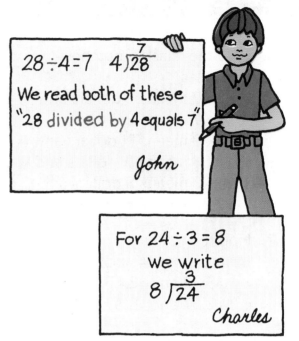

John made a poster to show a new way to write division exercises.

1. Show how to write the following division equation using the new way.

    $45 \div 5 = 9$

2. Explain what Charles did wrong on his paper.

3. Which letter—A, B, or C—represents the quotient?

4. Fran wanted to find how many pages she would need for 78 doll pictures, if she put 3 pictures on each page. She wrote her work like this.
    - A How many threes did Fran subtract the first time? the second time? the third time?
    - B How many threes did she subtract altogether? What is the quotient? How many pages does she need for the doll pictures?

Using the Ideas

1. Write each of these exercises using the new method.

   A $12 \div 4 = 3$ B $14 \div 2 = 7$ C $48 \div 6 = 8$ D $63 \div 9 = 7$

2. Find the quotients.

   A $2\overline{)6}$ B $2\overline{)14}$ C $8\overline{)24}$ D $8\overline{)40}$ E $8\overline{)56}$

   F $5\overline{)25}$ G $4\overline{)16}$ H $4\overline{)32}$ I $6\overline{)36}$ J $7\overline{)56}$

3. Find the quotients.

    A
    $$2\overline{)46}$$
    $$\underline{20} \quad (10)$$
    $$26$$
    $$\underline{20} \quad (10)$$
    $$6$$
    $$\underline{6} \quad (3)$$
    $$0$$

    B
    $$5\overline{)85}$$
    $$\underline{50} \quad (10)$$
    $$35$$
    $$\underline{35} \quad (7)$$
    $$0$$

    C
    $$4\overline{)144}$$
    $$\underline{80} \quad (20)$$
    $$64$$
    $$\underline{40} \quad (10)$$
    $$24$$
    $$\underline{24} \quad (6)$$
    $$0$$

    D
    $$3\overline{)114}$$
    $$\underline{90} \quad (30)$$
    $$24$$
    $$\underline{24} \quad (8)$$
    $$0$$

4. Copy each exercise and give the missing numbers.

    A
    $$3\overline{)42}$$
    $$\underline{30} \quad (?)$$
    $$12$$
    $$\underline{12} \quad (?)$$
    $$0$$

    B
    $$6\overline{)78}$$
    $$\underline{60} \quad (?)$$
    $$18$$
    $$\underline{18} \quad (?)$$
    $$0$$

    C
    $$4\overline{)92}$$
    $$\underline{80} \quad (?)$$
    $$12$$
    $$\underline{12} \quad (?)$$
    $$0$$

    D
    $$7\overline{)105}$$
    $$\underline{70} \quad (?)$$
    $$35$$
    $$\underline{35} \quad (?)$$
    $$0$$

5. Find the quotients.

   A $2\overline{)34}$ B $5\overline{)65}$ C $3\overline{)48}$ D $6\overline{)72}$ E $4\overline{)96}$ F $7\overline{)84}$

   G $8\overline{)96}$ H $9\overline{)108}$ I $4\overline{)68}$ J $7\overline{)98}$ K $6\overline{)90}$ L $3\overline{)51}$

More practice, page A-36, Set 48

● *Let's use division to solve problems.*

Investigating the Ideas

Divide sets of counters to find the answers to these problems.

1.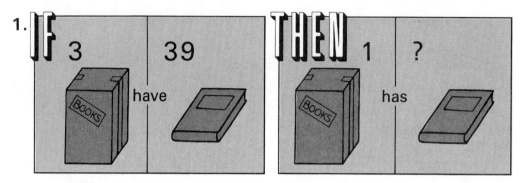

2.

 Can you write and complete a division problem that would help you answer the questions above?

Discussing the Ideas

1. There is a total of 54 washers in 3 boxes. Write and solve a division problem that will tell you how many washers are in each box. Why is division used to solve the problem?

2. Can you make up a problem like those above and explain how to solve it by using division?

Using the Ideas

Solve the problems.

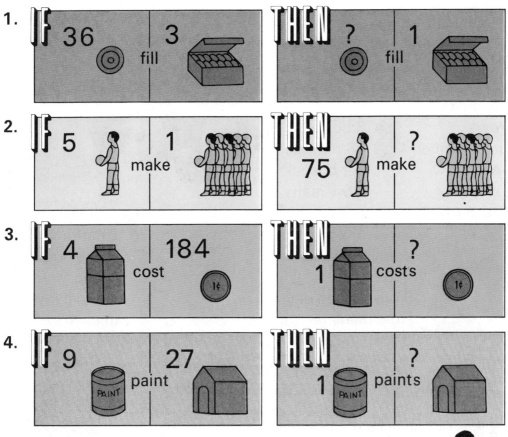

think

There are 3 stacks of checkers on a table. 6 checkers were removed from 1 stack and divided equally between the other 2 stacks. Then all the stacks had the same number. At the beginning, how many more checkers were in the tall stack than in the shorter ones?

Solving Short Stories

1 78 children.
9 more came later.
How many children in all?

2 162 children.
Same number of children in each of 6 groups.
How many in each group?

3 82 children. 37 boys. How many girls?

4 315 chairs.
7 rows.
Same number in each row.
How many in each row?

5 6 bushels.
48 pounds per bushel.
How many pounds in all?

6 Satellite makes 558 orbits.
9 orbits each day.
How many days?

7 108 eggs.
9 cartons.
Same number in each carton.
How many in each carton?

8 7 days in a week. 364 days. How many weeks?

9 112 ounces are 7 pounds.
How many ounces in 1 pound?

10 5280 feet in a mile.
How many feet in 2 miles?

11 87 cents for 3 pints of ice cream.
How much per pint?

12 4 quarts in a gallon.
96 cents per gallon.
How much per quart?

★ **13** 2 pints in a quart.
4 quarts in a gallon.
$1.92 (192 cents) per gallon.
How much per pint?

TIME

1. 4 o'clock now. Sleep for 7 hours. What time will it be?

2. Walk a mile in 23 minutes. Run a mile in 9 minutes. How much quicker to run?

3. 60 minutes in 1 hour. 8 hours. How many minutes?

4. Machine runs 424 minutes a day. Makes one thing each 8 minutes. How many things?

5. 1 day has 24 hours. 9 days. How many hours?

6. 126 days make 9 fortnights. 1 fortnight is how many days?

7. Total sleep, 736 hours. 8 hours each day. How many days?

8. 658 days. How many weeks?

9. Turbojet goes 577 mph (miles per hour). Gas-engine plane goes 334 mph. How much faster is the turbojet?

10. 420 seconds. Same as 7 minutes. How many seconds in a minute?

Animals, trees, birds, and insects grow old. Exercises 11 through 18 tell how old they sometimes grow.

11. An old cat: 15 years old. An old turtle: 10 times as old. How old is an old turtle?

12. An old rabbit: 6 years old. An old goose: 11 times as old. How old is an old goose?

13. An old bear: 35 years old. An old camel: 28 years old. How much older is an old bear than an old camel?

14. An old reindeer: 12 years old. An old whale: 6 times as old. How old is an old whale?

15. Spruce tree: 243 years old. Lives 339 more years. How old?

16. An old eagle: 46 years old. An old elm tree: 7 times as old. How old is an old elm tree?

17. An old elephant: 61 years old. An old cow: 24 years old. How much older is an old elephant than an old cow?

18. An old butterfly: 8 weeks old. An old housefly: 6 weeks old. How many **days** older is the old butterfly?

● *Is the remainder always 0?*

Discussing the Ideas

Sara had 51 photographs to put into her new photo book. She could put 4 pictures on a page. She decided to use division to find how many pages she would need.

1. When Sara divided 51 by 4, what did she get for the quotient?

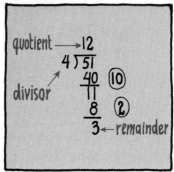

2. After Sara used 12 pages of the book, how many photos would she have left?

3. Notice that 3 is called the **remainder** and 4 the **divisor**.
   A If 4 is the divisor, could you have a remainder of 4? Explain.
   B Could you have a remainder greater than the divisor?

4. Explain this rule.

   > Always carry out the dividing until the remainder is less than the divisor.

5. Give the quotient, divisor, and remainder for each example. Is the remainder always less than the divisor?

```
  A      6        B      3        C     27         D     45
      4)26            7)25            5)136            3)135
        24              21              100              120
       ───             ───             ───              ───
         2               4               36               15
                                         35               15
                                        ───              ───
                                          1                0
```

Using the Ideas

1. Find the quotients and remainders.

   A 3)28　　B 4)39　　C 5)42　　D 6)45　　E 7)59

   F 8)78　　G 9)71　　H 4)46　　I 3)98　　J 7)166

   K 4)237　　L 8)337　　M 3)181　　N 9)376　　O 5)364

2. Susan had 57 pictures for her new album. She put 6 pictures on each page.

   A How many full pages could she get?
   B How many pictures would be left over for the last page?

3. Jim has 221 Indian-head pennies. His coin book has room for 9 pennies on each page.
   A How many full pages could he get?
   B How many pennies would be left over for the last page?

★ 4. Can you write and solve a division problem that will help you answer each question?

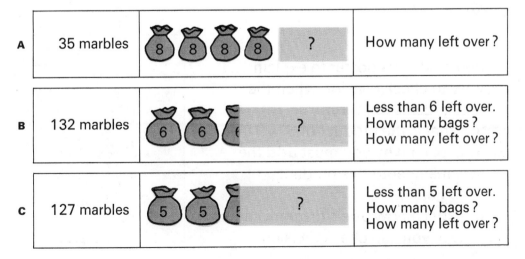

A	35 marbles	8 8 8 8 ?	How many left over?
B	132 marbles	6 6 6 ?	Less than 6 left over. How many bags? How many left over?
C	127 marbles	5 5 5 ?	Less than 5 left over. How many bags? How many left over?

More practice, page A-36, Set 49

● How can answers in division be checked?

Investigating the Ideas

Find each product.

A 27   B 15   C 44
  × 4   × 6   × 3

D 34   E 32   F 36
  × 7   × 5   × 3

Cynthia
1. 238 ÷ 7 = 34
2. 160 ÷ 5 = 32
3. 108 ÷ 4 = 24
4. 108 ÷ 3 = 34
5. 90 ÷ 6 = 15
6. 132 ÷ 3 = 42

 Can you use your answers to the exercises above to help you grade this paper?

Discussing the Ideas

1. Explain how Patty can use multiplication to see if she has found the quotient for 48 ÷ 4.

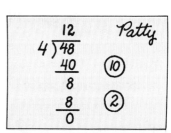

2. Joe made this poster to explain how to check division when the remainder is not zero.
    A What is the product of 3 × 14?
    B Explain why you must **add** the remainder to this product to get 44.
    C Explain how Joe's diagram shows a way you can check division.

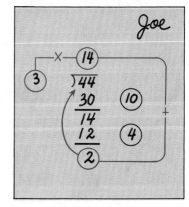

Using the Ideas

1. Find the quotients and remainders. Then check your answers.

   A 4)25   B 5)27   C 3)10   D 6)39   E 2)19

   F 5)37   G 6)29   H 7)65   I 4)30   J 3)29

   K 8)206   L 6)154   M 5)276   N 7)185   O 6)134

2. In a given part of this exercise, each bag contains the same number of marbles. Use division to help you find the answer. Check your work.

A	161 marbles	7 bags (?)	How many in each bag?
B	43 marbles	4 bags (?) with 3 dots	3 left over. How many in each bag?
C	39 marbles	bags of 7, ?	4 left over. How many bags?
D	27 marbles	4 bags (?) with 3 dots	3 left over. How many in each bag?
E	44 marbles	?, bag of 5	4 left over. How many bags?
F	60 marbles	bag of 9, ?	Less than 9 left over. How many bags? How many left over?
G	184 marbles	bag of 8, ?	Less than 8 left over. How many bags? How many left over?

More practice, page A-37, Set 50

Reviewing the Ideas

1. Write a division equation to answer each question.

   A  How many sets of 3 are in a set of 21?

   B 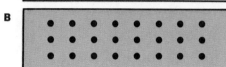 If we put 24 dots into 3 sets of the same number, how many dots are in each set?

   C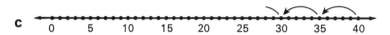
   Starting at 40, how many jumps of 5 does it take to get to 0?

   D   42   35   28
       −7   −7   −7   . . .   Starting with 42, how many times
       ──   ──   ──   do we subtract 7 to get 0?
       35   28

   E $n \times 9 = 54$ What number times 9 gives 54?

2. Find the quotients.
   Since $11 \times 23 = 253$, we know that $253 \div 23 = n$.
                                                $253 \div 11 = n$.

3. Find the quotients.

   A $64 \div 8$ I $49 \div 7$ Q $81 \div 9$
   B $28 \div 4$ J $40 \div 8$ R $18 \div 2$
   C $20 \div 5$ K $48 \div 6$ S $25 \div 5$
   D $18 \div 6$ L $54 \div 6$ T $45 \div 5$
   E $27 \div 9$ M $24 \div 6$ U $30 \div 5$
   F $45 \div 5$ N $32 \div 8$ V $36 \div 9$
   G $42 \div 7$ O $56 \div 8$ W $16 \div 4$
   H $36 \div 6$ P $72 \div 9$ X $35 \div 7$

think

Brian's father weighs 100 pounds more than Brian. Together they weigh 240 pounds. How much does Brian weigh?

4. Find the quotients.

    A 5)315   B 6)264   C 3)228   D 4)372   E 2)168

    F 4)300   G 7)700   H 6)420   I 8)272   J 9)558

5. Find the quotients and the remainders.

    A 3)128   B 5)342   C 6)415   D 4)321   E 2)101

    F 8)327   G 9)576   H 3)194   I 7)627   J 6)245

6.	258 marbles	🎒🎒🎒🎒🎒🎒	How many in each bag?
7.	175 marbles	🎒🎒 ?	How many bags?
8.	?	🎒🎒🎒🎒🎒🎒🎒	How many marbles?

9. Jan had 35 balloons for her party. There were 8 children at the party. Each child got the same number of balloons, and there were 3 left over. How many balloons did each child get?

10. Jim had 50 cents. Table tennis balls cost 9 cents each.
    A How many could he buy? How much money is left over?
    B If he bought only 3 balls, how much money would he have left?

11. Sara had 75 cents when she went shopping.
    A How many pencils could she buy if they were 6 cents each? How much would she have left?
    B If pencils were 9 cents each, how many could she buy? How much money would she have left?

Keeping in Touch with Addition, Subtraction, Multiplication, Division, Story problems

1. Find the sums, products, quotients, and differences.

   A. 94 + 39
   B. 68 × 3
   C. 78 − 52
   D. 81 + 79
   E. 63 × 5
   F. 79 × 6
   G. 27 + 88
   H. 56 × 8
   I. 65 + 99
   J. 93 × 7
   K. 142 − 80
   L. 125 − 52
   M. 65 − 23
   N. 350 ÷ 7
   O. 350 ÷ 5
   P. 540 ÷ 9
   Q. 120 ÷ 3

2. Tell what operation (+, −, ×, ÷) you think of for:
   A. putting 2 sets together and finding the total number.
   B. finding how many are left after some have been taken away.
   C. finding how many sets of a certain size we get from a set.
   D. finding how many in a certain number of rows of the same number.
   E. finding how many more one set has than another.
   F. finding how many ways we can pair objects in 2 sets.
   G. finding how many rows when we put a set into rows having the same number.

3. Solve the equations.
   A. $n + 6 = 11$
   B. $8 + n = 15$
   C. $3 \times n = 18$
   D. $n \times 8 = 24$
   E. $50 \div 5 = n$
   F. $8 - n = 6$
   G. $10 \div n = 2$
   H. $42 \div n = 6$
   I. $n - 8 = 6$
   J. $n \div 6 = 5$
   K. $18 + n = 24$

think

A train that is 1 mile long is traveling 1 mile each 3 minutes. How long does it take this train to pass through a 2-mile tunnel?

Solving Story Problems

1. Shirt costs $4.89.
Jacket costs $11.56.
How much less is the shirt?

2. 182 miles on Friday.
496 miles on Saturday.
527 miles on Sunday.
How far in the three days?

3. 128 pickles in a barrel.
4 barrels in storeroom.
How many pickles in all?

4. Peaches: 12 cents each.
Pears: 16 cents each.
Michael bought 7 peaches
and 9 pears. How much
did he spend?

5. 138 baseball cards. 6 per package. How many packages?

6. 513 pansies in cartons.
9 cartons.
How many pansies
per carton?

7. Use 9 gallons
of gasoline to go
207 miles. How many
miles traveled
on each gallon?

8. 184 tickets. 8 bundles
of the same size.
How many in each bundle?

9. Sixty-five 8-cent stamps.
Nine 11-cent stamps.
How much change from a 10-dollar bill?

10. Forty-eight 8-cent stamps.
Sixteen 10-cent stamps.
One hundred ninety-six 2-cent stamps.
How much change from a 10-dollar bill?

Eating at the Restaurant

Kay and her mother went to a restaurant in the city.

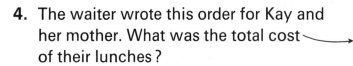

1. There were 36 tables in the restaurant. There were 4 chairs at each table. How many chairs were there?

2. There were 9 waiters. How many tables might be assigned to each waiter?

3. The waiters used 1 pitcher of water for each 4 people. Kay counted 56 people. How many pitchers of water did they need?

4. The waiter wrote this order for Kay and her mother. What was the total cost of their lunches?

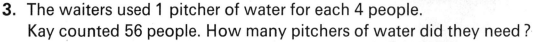

5. Kay's mother gave the waiter $4.50. How much change did she get back?

6. Their lunch cost about $4.00. Kay's mother decided to tip the waiter 15 cents for each dollar that the lunch cost. How much extra did she give the waiter for his good service?

7. Kay saw this sign in the window.
   - A How many weeks has the restaurant been open without closing?
   - ★ B How many hours has it been open without closing?

You are invited to explore

ACTIVITY CARD 17
Page 317

Mathematical Activities

How to Use the Activity Cards

Do you like to explore things for yourself? These Activity Cards will give you some exciting experiences with mathematics. Each card presents a different idea for you to explore. Often you will find that a card will give you ideas for additional activities on your own.

ACTIVITY CARD 1

A is the smallest **square** you can form on the geoboard with one rubber band.

B is the largest square you can form on the geoboard with one rubber band.

Six more squares of different sizes can be formed.

How many of them can you find and draw on dot paper?

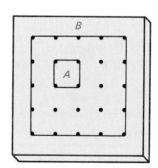

ACTIVITY CARD 2

Rubber band A encloses an **area** of 1 square.

Rubber band B encloses an area of 2 squares.

The area of C is 4 squares.

Can you show a **square** or a **rectangle** that has an area of 3? 5? 6? 7? 8? 9? 10? 11? 12? 13? 14? 15? 16?

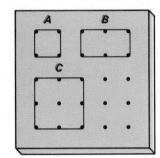

ACTIVITY CARD 3

Mark 3 points, A, B, and C, on a sheet of paper. Use a black and a red loop of string. In the figure, A and C are inside only one loop. B is inside both loops.

How many ways can you place the string so that

▶ each dot is inside some loop
  and
  exactly one dot is inside both loops?

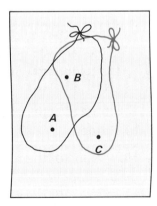

ACTIVITY CARD 4

A and B are different in **shape** and **color**.
B and D are different in **shape** and **size**.

A

▶ In how many ways are A and C different?

B

▶ In how many ways are B and C different?

C

▶ Can you color and cut out a figure that is different from D in 3 ways?

D

ACTIVITY CARD 5

If you toss a penny 10 times, how many heads do you think you will get?
Try it.

Guess how many heads you will get in 100 tosses.
Now try it.

Can you predict about how many heads you would get in 1000 tosses?

ACTIVITY CARD 6

How many "SQUARES"* can you find in your class?

(Figure out an easy way to measure to see if a person is a "square.")

*Someone who can fit exactly inside a square

ACTIVITY CARD 7

Put a marker on start. Flip two pennies.
Move one space left if neither penny is a head.
Move one space right if 1 or 2 heads show.
If you keep doing this, which do you think you will reach first, the chicken or the egg?
Try it.

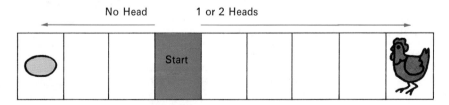

ACTIVITY CARD 8

Can you stand on a sidewalk and estimate where you will be after 100 normal steps?
Try it.

How many of your 100-step distances do you think it would take to go one mile?

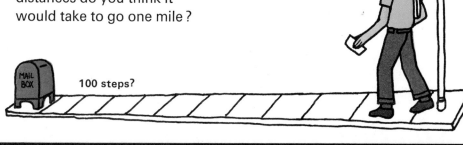

ACTIVITY CARD 9

Suppose there are 3 empty seats in your classroom. Two new children join your class.

Can you find how many different ways your teacher could give them seats?

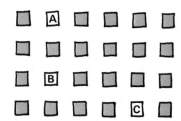

Jim

Pam

ACTIVITY CARD 10

Trace these shapes and cut out ten of each.

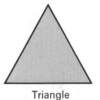
Triangle

Quadrilateral

Pentagon

Hexagon

Which of these shapes could be used to tile a floor? (The tiles must not overlap or have any space between them.)

Show each answer by pasting the ten shapes on a paper as if you were starting to tile the floor.

ACTIVITY CARD 11

Here is a way to make a 4-by-4 magic square.

Number a 4-by-4 square consecutively as in square **A**, starting with 3.

Exchange positions of the pairs of numerals connected by the arrows to get square **B**.

What is the magic sum in each row, column, and diagonal of square **B**?

Can you make your own magic square by starting with a different number?

A

3	4	5	6
7	8	9	10
11	12	13	14
15	16	17	18

B

18	4	5	15
7	13	12	10
11	9	8	14
6	16	17	3

ACTIVITY CARD 12

How many letters are on the front page of your newspaper?

Can you find a way to estimate this number of letters without actually counting them all?

ACTIVITY CARD 13

How close can you come to finding length by counting your steps?

(Practice taking a 2-foot or 3-foot step. Then choose some distances to measure and make a table like the one shown.)

Distance to measure	By counting steps	By using a ruler or tape	Difference
Room length			
Room width			

ACTIVITY CARD 14

Here is the way Richard used the telephone dial to find the sum of the letters in his name.

R I C H A R D
| | | | | | |
7 + 4 + 2 + 4 + 2 + 7 + 3 = 29

Can you find the sum for your name in this way?
Which one of your classmates has the largest sum for his name?
(If a name has a **Q** or **Z**, count it as zero.)

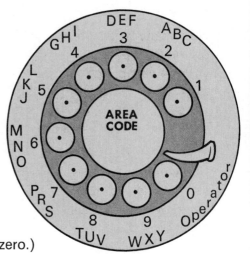

ACTIVITY CARD 15

Fold a piece of paper.

Make a cut that starts on the fold and ends on the fold.

Unfold the piece you cut out. It will be **symmetric** about the fold line.

Can you use this method to make a square?
a rectangle? a heart? a triangle? a pumpkin? a letter of the alphabet? a rocket? a butterfly? a funny person?

ACTIVITY CARD 16

Can you figure out a way to use a rectangular sheet of graph paper to make a picture of Merfel the Mule **just like this one only larger?**

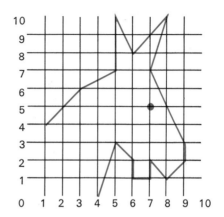

ACTIVITY CARD 17

~~1~~ 2 3 4 5 ~~6~~ 7 8 9

Play this game with a classmate.
Write the numerals 1 through 9 on a piece of paper.
Toss two dice. Mark out the sum you toss. For example, if you toss a 5 and a 2, you may mark out any combination of digits that totals 7. Continue tossing the dice until you can no longer mark out the sum you toss from the remaining digits.
Your score is the sum of the digits that remain.
The person with the lower total wins.

Appendix

More Practice	A-1
Books to Explore	A-38
Glossary	A-41
Tables of Measures	A-44
Index	A-45

More Practice

Set 1 *For use with page 13*

Give the length of each object to the nearest inch.

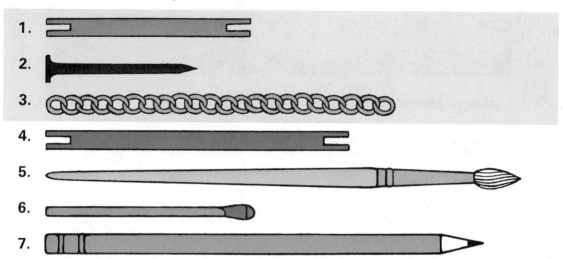

Give the length of each object to the nearest centimeter.

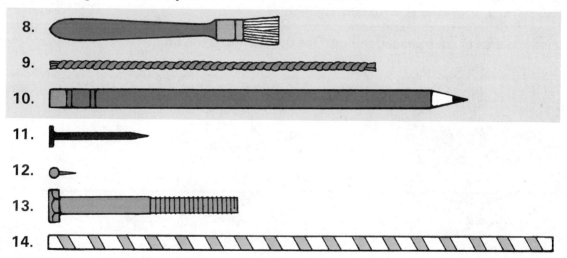

Reflected answers, Set 1: 1. 2˙ 2. 2˙ 3. 4˙ 8. 6˙ 9. 9˙ 10. 11

A-1

Set 2 — For use with page 15

Measure each object to the nearest half inch.

1.
2.
3.
4.
5.
6.
7.

Reflected answers, Set 2: 1. 2, 2. 5, 3. 2½, 3. 2

Set 3 — For use with page 19

Find the area of each shaded region. Use the square as the unit.

1.    2.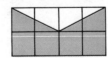

3.   4.

5.   6.

Reflected answers, Set 3: 1. 2, 2. 5, ʼ ⱱ

Set 4   *For use with page 23*

Find each length to the nearest fourth inch.

1.
2.
3.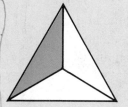
4.

Give the fraction that tells what part of each region is shaded.

5.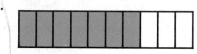
6.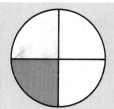
7.
8.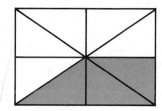
9.
10.

Solve each short story.

11. 4 candy bars.
    Gave away $\frac{1}{2}$ of them.
    How many left?

12. 6 children.
    $\frac{1}{3}$ of them are girls.
    How many are girls?

13. 12 cookies.
    Dropped $\frac{1}{4}$ of them.
    Dropped how many?

14. 12 dimes.
    Spent $\frac{1}{3}$ of them.
    Spent how many?

Reflected answers, Set 4:   1. 3, 2. 2$\frac{3}{4}$, 5. $\frac{3}{3}$, 6. $\frac{1}{4}$, 11. 2, 12. 2

A-3

Set 5 — For use with page 31

Write the 2-digit numeral for each of these.

1. 2 tens and 5
2. 3 tens and 6
3. 1 ten and 2
4. 4 tens and 0
5. 1 ten and 5
6. 7 tens and 3
7. 8 tens and 7
8. 4 tens and 9
9. 5 tens and 5
10. 9 tens and 0
11. 2 tens and 8
12. 8 tens and 1

Give the correct digit for each ▦.

13. 21 means 2 tens and ▦.
14. 37 means 3 tens and ▦.
15. 89 means ▦ tens and 9.
16. 75 means ▦ tens and 5.
17. 40 means ▦ tens and 0.
18. 63 means 6 tens and ▦.
19. 92 means 9 tens and ▦.
20. 38 means ▦ tens and 8.
21. 56 means ▦ tens and 6.
22. 10 means 1 ten and ▦.

Reflected answers, Set 5: 1. 25, 2. 36, 5. 15, 6. 73, 9. 55, 10. 90, 13. 1, 14. 7, 18. 3, 19. 2.

Set 6 — For use with page 33

Write the word name for each of these.

1. 5 tens and 6
2. 3 tens and 5
3. 7 tens and 0
4. 8 tens and 1
5. 1 ten and 9
6. 4 tens and 7
7. 2 tens and 8
8. 6 tens and 2
9. 1 ten and 4
10. 9 tens and 9
11. 8 tens and 6
12. 2 tens and 9

Copy each row. Give the missing numbers.

13. 7, 8, 9, 10, ?, ?, ?, ?, 15, 16
14. 25, 26, 27, 28, ?, ?, ?, ?, 33, 34
15. ?, ?, ?, ?, 40, 41, 42, 43
16. 80, 81, 82, 83, ?, ?, ?, ?, 88, 89
17. 2, 4, 6, 8, ?, ?, ?, ?, 18, 20
18. 5, 10, ?, ?, ?, ?, 35, 40

Reflected answers, Set 6: 1. fifty-six, 2. thirty-five, 5. nineteen, 9. fourteen, 10. ninety-nine, 13. 11, 12, 13, 14, 14. 29, 30, 31, 32, 15. 36, 37, 38, 39.

A-4

Set 7   For use with page 37

Write the numeral. (*h* stands for *h*undreds and *t* for *t*ens.)

1. 5*h*, 3*t*, and 4
2. 6*h*, 5*t*, and 1
3. 4*h*, 1*t*, and 9
4. 1*h*, 2*t*, and 5
5. 3*h*, 8*t*, and 2
6. 7*h*, 9*t*, and 0
7. 9*h*, 0*t*, and 3
8. 8*h*, 0*t*, and 8
9. 2*h*, 5*t*, and 6
10. 3*h*, 8*t*, and 2
11. 5*h*, 7*t*, and 0
12. 1*h*, 3*t*, and 9
13. 7*h*, 9*t*, and 6
14. 4*h*, 0*t*, and 0
15. 9*h*, 1*t*, and 5
16. 8*h*, 4*t*, and 8
17. 2*h*, 2*t*, and 2
18. 6*h*, 5*t*, and 3

Give the missing digit.

19. 226 means 2 hundreds, ____ tens, and 6 ones.
20. 384 means 3 hundreds, 8 tens, and ____ ones.
21. 179 means ____ hundreds, 7 tens, and 9 ones.
22. 838 means ____ hundreds, 3 tens, and 8 ones.
23. 915 means 9 hundreds, ____ tens, and 5 ones.
24. 475 means 4 hundreds, 7 tens, and ____ ones.
25. 508 means 5 hundreds, ____ tens, and 8 ones.
26. 657 means ____ hundreds, 5 tens, and 7 ones.
27. 830 means 8 hundreds, 3 tens, and ____ ones.
28. 100 means ____ hundreds, 0 tens, and 0 ones.

Write the numeral for each part.

29. one hundred sixty-six
30. two hundred seventeen
31. five hundred ninety-nine
32. three hundred thirty-seven
33. eight hundred two
34. four hundred twenty-two
35. nine hundred fifty
36. seven hundred
37. one hundred ninety-three
38. nine hundred twelve
39. five hundred sixty-one
40. seven hundred forty-five
41. two hundred six
42. three hundred ninety-nine

Reflected answers, Set 7: 1. 534, 2. 651, 7. 903, 8. 808, 13. 796, 14. 400, 19. 2, 20. 4, 21. 1, 22. 8, 23. 1, 29. 166, 30. 217, 31. 599, 36. 700, 37. 193, 38. 912

A-5

Set 8 — For use with page 41

Write the 4-digit numeral for each of these. (*th* stands for *thousands*, *h* stands for *hundreds*, and *t* stands for *tens*.)

1. 5*th*, 3*h*, 8*t*, 5
2. 7*th*, 8*h*, 2*t*, 4
3. 9*th*, 6*h*, 1*t*, 3
4. 1*th*, 0*h*, 3*t*, 9
5. 6*th*, 4*h*, 6*t*, 7
6. 2*th*, 4*h*, 8*t*, 8
7. 3*th*, 5*h*, 9*t*, 1
8. 8*th*, 9*h*, 0*t*, 6
9. 4*th*, 9*h*, 0*t*, 0
10. 8*th*, 1*h*, 7*t*, 2
11. 2*th*, 8*h*, 1*t*, 3
12. 9*th*, 9*h*, 9*t*, 9

Find the missing digit for each of these.

13. 4872 means 4 thousands, ____ hundreds, 7 tens, 2 ones.
14. 5396 means 3 hundreds, 6 ones, 9 tens, ____ thousands.
15. 6003 means ____ hundreds, 0 tens, 6 thousands, 3 ones.
16. 9218 means 1 ten, ____ thousands, 2 hundreds, 8 ones.

Set 9 — For use with page 43

Which of the two numbers is greater?

1. 9 or 7
2. 12 or 20
3. 35 or 25
4. 126 or 226
5. 450 or 460
6. 796 or 790
7. 999 or 1999
8. 4921 or 4931
9. 8890 or 8889

Place the correct sign (< or >) between each pair of numbers.

10. 65 ⬤ 55
11. 34 ⬤ 64
12. 97 ⬤ 79
13. 376 ⬤ 476
14. 581 ⬤ 571
15. 783 ⬤ 873
16. 575 ⬤ 585
17. 3092 ⬤ 3192
18. 6426 ⬤ 5426
19. 2361 ⬤ 2351
20. 8805 ⬤ 8905
21. 4223 ⬤ 4333

| Set 10 | *For use with page 55* |

Find the sums.

1. 0 +1	2. 3 +5	3. 0 +3	4. 1 +2	5. 0 +4	6. 1 +5	7. 4 +2	8. 6 +2
9. 2 +4	10. 0 +2	11. 7 +0	12. 6 +1	13. 0 +6	14. 1 +7	15. 2 +1	16. 1 +3
17. 2 +6	18. 2 +5	19. 4 +4	20. 1 +8	21. 2 +0	22. 0 +5	23. 5 +3	24. 2 +2
25. 3 +4	26. 1 +1	27. 0 +8	28. 1 +6	29. 3 +6	30. 5 +2	31. 8 +1	32. 3 +0
33. 6 +3	34. 7 +2	35. 8 +0	36. 1 +4	37. 3 +2	38. 4 +5	39. 6 +4	40. 2 +3

Reflected answers, Set 10: 1. 1, 2. 8, 3. 3, 4. 3, 5. 4, 6. 6, 7. 6, 8. 8, 9. 6, 10. 2, 11. 7, 12. 7, 13. 6, 14. 8, 15. 3, 16. 4

| Set 11 | *For use with page 57* |

Find the missing addends.

1. ___ + 3 = 7   3. ___ + 2 = 5   5. ___ + 4 = 10
2. ___ + 5 = 9   4. ___ + 1 = 8   6. ___ + 5 = 8

Find the differences.

7. 3 −0	8. 5 −1	9. 8 −7	10. 4 −4	11. 1 −0	12. 2 −2	13. 2 −1	14. 8 −2
15. 9 −4	16. 9 −7	17. 6 −5	18. 7 −3	19. 10 −8	20. 7 −7	21. 6 −4	22. 6 −3

Reflected answers, Set 11: 1. 4, 3. 3, 5. 6, 2. 4, 4. 7, 6. 3, 7. 3, 8. 4, 9. 1, 10. 0, 11. 1, 12. 0, 13. 1, 14. 6

Set 12 — For use with page 59

Find the differences.

1. 9 −5	**2.** 6 −5	**3.** 9 −3	**4.** 9 −6	**5.** 9 −8	**6.** 8 −3	**7.** 8 −6	**8.** 7 −5
9. 9 −9	**10.** 8 −0	**11.** 7 −2	**12.** 7 −4	**13.** 10 −9	**14.** 10 −7	**15.** 10 −4	**16.** 10 −5

Solve each story problem.

17. Bill had 10 marbles. He gave John 6 of them. How many marbles does Bill have left?
18. Betty is 9 years old. Her sister is 5 years old. How much older is Betty?
19. Ann has 8 cents. Dick has 3 cents less. How many cents does Dick have?

Reflected answers, Set 12: [reflected]

Set 13 — For use with page 63

Find the sums.

1. 4 1 +2	**2.** 2 4 +3	**3.** 1 5 +2	**4.** 2 5 +3	**5.** 3 1 +0	**6.** 6 2 +2	**7.** 5 1 +2	**8.** 3 1 +6
9. 6 4 +3	**10.** 6 7 +3	**11.** 9 8 +1	**12.** 2 8 +7	**13.** 5 4 +5	**14.** 6 8 +2	**15.** 1 5 +9	**16.** 8 9 +2

17. 3 + 7 + 3 + 2
18. 7 + 5 + 3 + 2
19. 2 + 3 + 8 + 7
20. 1 + 7 + 9 + 5
21. 8 + 5 + 2 + 6
22. 4 + 9 + 6 + 8

Reflected answers, Set 13: [reflected]

A-8

Set 14 *For use with page 65*

Find the sums.

1. 9 +2	2. 3 +7	3. 6 +5	4. 8 +4	5. 6 +6	6. 4 +9	7. 1 +9	8. 6 +4
9. 9 +3	10. 8 +2	11. 2 +9	12. 5 +7	13. 6 +8	14. 8 +7	15. 9 +8	16. 8 +3
17. 7 +7	18. 9 +5	19. 5 +6	20. 3 +9	21. 7 +8	22. 8 +9	23. 9 +7	24. 4 +7
25. 8 +8	26. 7 +6	27. 3 +8	28. 5 +8	29. 6 +9	30. 4 +8	31. 6 +7	32. 9 +6
33. 8 +6	34. 9 +4	35. 5 +9	36. 9 +9	37. 7 +9	38. 8 +5	39. 7 +5	40. 7 +4
41. 6 4 +3	42. 7 3 +6	43. 9 1 +8	44. 2 8 +7	45. 5 5 +4	46. 6 8 +2	47. 7 2 +3	48. 1 5 +9

49. 2 + 5 + 3
50. 6 + 2 + 4
51. 3 + 2 + 9
52. 3 + 4 + 6
53. 5 + 9 + 5
54. 6 + 7 + 4
55. 5 + 7 + 6
56. 9 + 6 + 3
57. 8 + 7 + 2
58. 9 + 4 + 1 + 3
59. 6 + 2 + 4 + 8
60. 2 + 7 + 8 + 1
61. 8 + 4 + 2 + 6
62. 5 + 3 + 4 + 3
63. 1 + 0 + 6 + 8

Solve each short story.

64. 9 boy's bikes. 5 girl's bikes. How many bikes?

65. 8 robins. 9 blue jays. How many birds?

Set 15 For use with page 67

Find the differences.

1. 14 −7	2. 10 −6	3. 13 −8	4. 11 −8	5. 12 −9	6. 15 −7	7. 13 −4	8. 11 −2
9. 10 −3	10. 11 −7	11. 11 −9	12. 10 −8	13. 11 −4	14. 12 −3	15. 13 −5	16. 14 −6
17. 12 −7	18. 12 −8	19. 11 −6	20. 12 −5	21. 13 −7	22. 13 −9	23. 12 −6	24. 14 −5
25. 11 −5	26. 15 −6	27. 12 −4	28. 16 −8	29. 13 −6	30. 16 −7	31. 14 −8	32. 16 −9
33. 10 −1	34. 11 −3	35. 15 −8	36. 17 −9	37. 18 −9	38. 14 −9	39. 17 −8	40. 15 −9

41. 10 − 6
42. 15 − 9
43. 17 − 8
44. 15 − 6

45. 18 − 5
46. 17 − 7
47. 10 − 9
48. 16 − 5

49. 17 − 2
50. 19 − 8
51. 16 − 9
52. 10 − 7

53. 14 − 7
54. 12 − 7
55. 15 − 5
56. 19 − 8

Solve each short story.

57. 12 cats and dogs. 5 cats.
How many dogs?

58. 15 candy bars. Ate 7.
How many left?

59. 8 girls. 8 shoes.
How many more shoes needed?

60. 14 boys. 6 hats.
How many more hats needed?

61. 19 cents. Lost 7 cents.
How much left?

62. 18 children. 9 boys.
How many girls?

A-10

| Set 16 | *For use with page 95* |

Find the sums and differences.

1. 30 +46	2. 37 +22	3. 50 +27	4. 32 +56	5. 24 +63	6. 35 +54	7. 61 +18
8. 31 +24	9. 45 +31	10. 18 +51	11. 12 +86	12. 37 +61	13. 49 +40	14. 20 +68
15. 432 +264	16. 507 +462	17. 326 +453	18. 856 +133	19. 750 +108	20. 513 +264	
21. 460 −150	22. 463 −322	23. 872 −541	24. 768 −306	25. 952 −340	26. 638 −414	
27. 390 −240	28. 576 −215	29. 898 −344	30. 792 −410	31. 856 −313	32. 984 −753	

Reflected answers, Set 16: 1. 76, 2. 59, 3. 77, 4. 88, 5. 87, 6. 89, 7. 79, 21. 310, 22. 141, 23. 331, 24. 462, 25. 612, 26. 224.

| Set 17 | *For use with page 99* |

Find the sums.

1. 36 +92	2. 27 +81	3. 13 +94	4. 45 +84	5. 34 +74	6. 72 +65	7. 24 +85
8. 27 +34	9. 32 +59	10. 46 +58	11. 83 +39	12. 56 +68	13. 47 +56	14. 91 +67
15. 92 +88	16. 77 +38	17. 19 +46	18. 53 +98	19. 42 +78	20. 37 +95	21. 72 +97

Reflected answers, Set 17: 1. 128, 2. 108, 3. 107, 4. 129, 5. 108, 6. 137, 7. 109.

A-11

Set 18 — For use with page 101

Find the sums.

1. 37 + 8
2. 56 + 26
3. 87 + 36
4. 43 + 9
5. 68 + 24
6. 76 + 76
7. 58 + 4
8. 22 + 9
9. 68 + 27
10. 85 + 79
11. 67 + 5
12. 14 + 69
13. 58 + 69
14. 59 + 4
15. 26 + 6
16. 54 + 29
17. 45 + 98
18. 54 + 6
19. 37 + 38
20. 66 + 78
21. 38 + 4
22. 73 + 8
23. 69 + 25
24. 76 + 74
25. 86 + 5
26. 48 + 46
27. 75 + 99
28. 67 + 8

29. 536 + 587
30. 947 + 589
31. 654 + 878
32. 948 + 699

33. 677 + 388
34. 666 + 666
35. 385 + 769
36. 562 + 949

37. 326 + 468
38. 534 + 259
39. 421 + 309
40. 319 + 146
41. 573 + 152
42. 629 + 152

43. 467 + 368
44. 379 + 123
45. 378 + 147
46. 638 + 287
47. 456 + 269
48. 736 + 199

49. 629 + 153
50. 406 + 279
51. 384 + 455
52. 293 + 563
53. 425 + 395
54. 273 + 588

55. 763 + 578
56. 637 + 895
57. 666 + 444
58. 858 + 564
59. 647 + 886
60. 792 + 589

61. 890 + 469
62. 876 + 452
63. 893 + 486
64. 989 + 376
65. 793 + 537
66. 927 + 894

Set 19   *For use with page 107*

Find the differences.

1. 32 − 8
2. 51 − 9
3. 36 − 8
4. 22 − 5
5. 47 − 8
6. 83 − 9
7. 91 − 3

8. 73 − 27
9. 41 − 8
10. 66 − 7
11. 84 − 8
12. 40 − 2
13. 73 − 4
14. 57 − 8

15. 30 − 6
16. 42 − 15
17. 75 − 39
18. 60 − 54
19. 34 − 28
20. 26 − 17
21. 58 − 29

22. 53 − 18
23. 71 − 42
24. 84 − 29
25. 36 − 18
26. 97 − 49
27. 86 − 29
28. 57 − 28

29. 82 − 76
30. 92 − 66
31. 63 − 45
32. 40 − 29
33. 81 − 12
34. 65 − 39
35. 88 − 19

36. 123 − 29
37. 164 − 77
38. 186 − 98
39. 152 − 73
40. 285 − 37
41. 636 − 58
42. 347 − 59

43. 153 − 69
44. 174 − 89
45. 216 − 89
46. 741 − 53
47. 580 − 233
48. 634 − 126
49. 727 − 282

Find the missing numbers.

50. 56 − ▓ = 49
51. 42 − ▓ = 38
52. 73 − ▓ = 67
53. 139 − 4 = ▓
54. 537 − 6 = ▓

55. 713 − 9 = ▓
56. 675 − 7 = ▓
57. 416 − ▓ = 414
58. 537 − ▓ = 534
59. 879 − ▓ = 873

60. 657 − 8 = ▓
61. 43 − 39 = ▓
62. 254 − 48 = ▓
63. 163 − 159 = ▓
64. 442 − 39 = ▓

A-13

Set 20   *For use with page 111*

Find the sums and differences.

1. 72 +65	2. 90 +47	3. 63 −23	4. 57 +84	5. 96 −37	6. 453 +236	7. 613 −252

8. 513 +888	9. 416 −126	10. 780 +57	11. 32 +499	12. 653 −283	13. 716 +24	14. 876 +347

15. 64 +36	16. 79 −48	17. 83 +49	18. 117 −48	19. 356 +241	20. 572 −148	21. 629 −598

22. 823 +799	23. 962 −897	24. 430 +980	25. 651 −279	26. 365 +376	27. 888 −499	28. 777 +666

29. 671 −346	30. 537 +283	31. 528 −141	32. 628 +177	33. 430 +398	34. 732 −496	35. 974 −876

36. 516 +694	37. 876 +427	38. 765 −379	39. 657 −288	40. 837 +999	41. 940 −333	42. 777 +567

Solve each short story problem.

43. 16 on the bus. 14 more get on. How many on bus?	44. 59 blue marbles. 87 red marbles. How many marbles?
45. Bill weighs 90 pounds. Mary weighs 75. How much heavier is Bill?	46. 188 math books. 275 books in all. How many are not math books?

47. There are 243 boys in Lozano School. There are 186 girls. How many more boys than girls are in the school?

Reflected answers, Set 20: 1. 137, 2. 137, 3. 40, 4. 141, 5. 59, 6. 689, 7. 361, 8. 1401, 9. 290, 10. 837, 11. 531, 12. 370, 13. 740, 14. 1223, 15. 100, 16. 31, 17. 132, 18. 69, 19. 597, 20. 424, 21. 31, 43. 30, 44. 146

A-14

Set 21 *For use with page 115*

Find the total amounts.

1. $1.36
    7.54
	2. $4.25
	    3.57
		3. $3.44
		    5.39
			4. $5.65
			      .16
				5. $4.37
				    2.91
					6. $7.81
					    1.46

7. $8.34
    6.79
	8. $4.53
	      .88
		9. $5.64
		    8.97
			10. $3.75
			     7.29
				11. $1.18
				     9.84
					12. $9.26
					     6.97

13. $8.36
     1.26
	14. $3.58
	     5.15
		15. $9.46
		       .37
			16. $5.61
			     3.09
				17. $6.84
				     2.41
					18. $4.72
					     4.92

19. $9.67
      .43
	20. $10.83
	      3.59
		21. $14.47
		      2.97
			22. $11.63
			      4.88
				23. $8.64
				     5.99
					24. $16.27
					      3.86

Find the difference in the amounts.

25. $8.56
  −3.21
	26. $7.74
	  −4.53
		27. $4.83
		  −4.53
			28. $3.90
			  −1.30
				29. $6.38
				     .24
					30. $9.74
					    4.34

31. $6.72
    1.37
	32. $8.65
	    2.26
		33. $9.44
		    7.19
			34. $4.36
			     .84
				35. $7.29
				    4.93
					36. $6.14
					     .39

Solve each story problem.

37. Jane had $1.50. She spent 75 cents for some candy. How much did she have left?

38. George had $7.43. He spent $1.35 for a ball. He also spent $3.59 for a bat. How much did he have left?

Reflected answers, Set 21: 1. $8.90, 2. $7.82, 3. $8.83, 4. $5.81, 5. $7.28, 6. $9.27, 7. $15.13, 8. $5.41, 9. $14.61, 25. $5.35, 26. $3.21, 29. $6.62, 30. $14.08.

Set 22   *For use with page 117*

Find the sums and differences.

1. 30 57 +14	**2.** 25 39 +35	**3.** 18 27 +43	**4.** 63 29 +38	**5.** 56 74 +58	**6.** 64 39 +65	**7.** 52 87 +36
8. 671 −252	**9.** 362 −133	**10.** 437 −242	**11.** 612 −136	**12.** 906 −174	**13.** 751 −227	
14. 747 −288	**15.** 535 −237	**16.** 496 −369	**17.** 368 −179	**18.** 417 −254	**19.** 925 −287	

Copy each problem. Write the missing digits.

20. 15☐ +7☐7 893	**21.** 377 −26☐ 1☐9	**22.** ☐6☐ −138 722	**23.** 23☐ +6☐9 883	**24.** ☐32 −39☐ 140	**25.** 282 +☐☐4 876
26. 92☐ +☐78 11☐1	**27.** 94☐ +☐☐9 1284	**28.** ☐39 −46☐ 477	**29.** 8☐☐ +☐94 1574	**30.** 7☐3 −32☐ 434	**31.** 7☐7 +☐64 1301

Solve each short story problem.

32. 72 horses. Only 28 spotted. How many are not spotted?

33. $9.49 for shoes. $13.35 for dress. How much for both?

34. 128 cherry trees. 237 peach trees. How many trees?

35. 432 children. 257 boys. How many girls?

36. Lakers score 134 points. Knicks score 97. How many more points for the Lakers?

37. 246 Fords. 317 Chevrolets. 196 Plymouths. How many cars in the lot?

Reflected answers, Set 22: 1. 101, 2. 99, 3. 88, 4. 130, 5. 188, 6. 168, 7. 175, 8. 419, 9. 229, 10. 195, 11. 476, 12. 732, 13. 524.

Set 23 — For use with page 145

Solve each equation.

1. $6 \times 5 = (4 \times 5) + (n \times 5)$
2. $7 \times 5 = (3 \times 5) + (n \times 5)$
3. $8 \times 6 = (5 \times 6) + (n \times 6)$
4. $8 \times 5 = (n \times 5) + (6 \times 5)$
5. $6 \times 4 = (n \times 4) + (2 \times 4)$
6. $8 \times 3 = (6 \times 3) + (n \times 3)$
7. $7 \times 6 = (n \times 6) + (3 \times 6)$
8. $7 \times 7 = (2 \times 7) + (n \times 7)$
9. $8 \times 6 = (6 \times 6) + (n \times 6)$
10. $7 \times 4 = (4 \times 4) + (n \times 4)$
11. $9 \times 5 = (6 \times 5) + (n \times 5)$
12. $9 \times 6 = (n \times 6) + (5 \times 6)$

13. Since $(3 \times 5) + (2 \times 5) = 25$, we know that $5 \times 5 = n$.
14. Since $(3 \times 6) + (4 \times 6) = 42$, we know that $7 \times 6 = n$.

Reflected answers, Set 23: 1. 2, 2. 4, 3. 4, 7. 4, 8. 5, 13. 25

Set 24 — For use with page 147

Find the products.

1. 0×3
2. 2×1
3. 2×2
4. 2×3
5. 3×2
6. 3×4
7. 1×2
8. 3×0
9. 2×0
10. 5×2
11. 3×6
12. 1×0
13. 0×1
14. 2×5
15. 2×8
16. 0×2
17. 3×7
18. 6×3

Find the products.

19. 5×3
20. 3×8
21. 0×2
22. 4×3
23. 3×7
24. 2×8
25. 5×3
26. 2×6
27. 6×3
28. 3×9
29. 5×2
30. 0×0
31. 1×3
32. 8×2
33. 3×2
34. 9×1

Solve these equations.

35. $3 \times 0 = n$
36. $8 \times 2 = n$
37. $3 \times 5 = n$
38. $7 \times 1 = n$
39. $2 \times 9 = n$
40. $3 \times 7 = n$
41. $8 \times 1 = n$
42. $4 \times 3 = n$

Reflected answers, Set 24: 1. 0, 4. 6, 7. 2, 10. 10, 13. 0, 16. 0, 35. 0, 37. 15, 39. 18, 41. 8

Set 25 — For use with page 149

Find the products.

1. 4×5
2. 2×3
3. 0×5
4. 7×4
5. 2×8
6. 3×7
7. 8×2
8. 6×4
9. 2×5
10. 0×5
11. 8×5
12. 0×8
13. 2×7
14. 7×5
15. 0×0

16. 9×4
17. 2×0
18. 8×4
19. 0×0
20. 4×3
21. 8×3
22. 5×6
23. 9×2

24. 8×5
25. 5×3
26. 7×1
27. 5×9
28. 9×4
29. 8×1
30. 3×9
31. 4×6

Solve the equations.

32. $3 \times 5 = n$
33. $6 \times 5 = n$
34. $7 \times 4 = n$
35. $7 \times 5 = n$
36. $9 \times 4 = n$
37. $8 \times 5 = n$
38. $9 \times 5 = n$
39. $1 \times 9 = n$

Reflected answers, Set 25: 1. 20, 4. 28, 7. 16, 10. 0, 13. 14

Set 26 — For use with page 151

Find the products.

1. 5×4
2. 3×7
3. 4×8
4. 3×8
5. 7×4
6. 8×3
7. 3×4
8. 9×5
9. 4×0
10. 7×3
11. 1×5
12. 5×5
13. 6×4
14. 3×6
15. 5×1
16. 3×2
17. 1×3
18. 2×9
19. 0×5
20. 8×2
21. 5×7
22. 0×6
23. 2×7
24. 5×8
25. 4×4

Solve the equations.

26. $5 \times 8 = n$
27. $2 \times 0 = n$
28. $4 \times 7 = n$
29. $6 \times 4 = n$
30. $3 \times 9 = n$
31. $7 \times 1 = n$
32. $4 \times 5 = n$
33. $1 \times 8 = n$

Reflected answers, Set 26: 1. 20, 2. 21, 6. 24, 7. 12, 11. 5, 12. 25, 16. 6, 17. 3, 21. 35, 22. 0

A-18

Set 27 — For use with page 153

Find the products.

1. 6 × 7
2. 7 × 6
3. 7 × 0
4. 6 × 3
5. 4 × 8
6. 2 × 7
7. 5 × 0
8. 2 × 9
9. 6 × 1
10. 2 × 4
11. 8 × 6
12. 4 × 3
13. 8 × 7
14. 3 × 7
15. 0 × 8
16. 3 × 6
17. 5 × 5
18. 0 × 5
19. 3 × 8
20. 5 × 6
21. 9 × 6
22. 8 × 5
23. 5 × 1
24. 5 × 8
25. 9 × 7

Solve the equations.

26. 6 × 4 = n
27. 3 × 7 = n
28. 8 × 7 = n
29. 5 × 6 = n
30. 9 × 6 = n
31. 8 × 6 = n
32. 9 × 7 = n
33. 6 × 5 = n

Reflected answers, Set 27: 1. 42, 6. 14, 11. 48, 16. 18, 21. 54

Set 28 — For use with page 155

Find the products.

1. 2 × 1
2. 2 × 2
3. 3 × 3
4. 3 × 4
5. 4 × 0
6. 4 × 3
7. 5 × 5
8. 4 × 4
9. 6 × 6
10. 8 × 7
11. 4 × 5
12. 9 × 0
13. 1 × 8
14. 5 × 3
15. 6 × 5
16. 4 × 6
17. 7 × 6
18. 9 × 1
19. 1 × 3
20. 4 × 2
21. 0 × 0
22. 2 × 9
23. 6 × 3
24. 7 × 3
25. 3 × 7
26. 7 × 5
27. 7 × 2
28. 8 × 1
29. 8 × 3
30. 0 × 3
31. 8 × 5
32. 4 × 8
33. 6 × 9
34. 1 × 9
35. 7 × 9
36. 9 × 8
37. 2 × 8
38. 5 × 7
39. 6 × 8
40. 0 × 6
41. 5 × 4
42. 5 × 1
43. 1 × 5
44. 8 × 2
45. 6 × 4
46. 2 × 6
47. 6 × 7
48. 5 × 6
49. 3 × 8
50. 3 × 1
51. 2 × 5
52. 9 × 2
53. 8 × 4
54. 0 × 8
55. 3 × 9
56. 5 × 8
57. 9 × 6
58. 1 × 7
59. 8 × 6
60. 1 × 6
61. 0 × 1
62. 8 × 8
63. 5 × 9
64. 2 × 4
65. 9 × 4

Reflected answers, Set 28: 1. 2, 14. 15, 27. 14, 40. 0, 53. 32

Set 29   *For use with page 159*

Solve the equations.

1. $5 \times 3 = n$
2. $2 \times 6 = n$
3. $4 \times 8 = n$
4. $6 \times 8 = n$
5. $9 \times 7 = n$
6. $8 \times 3 = n$
7. $2 \times 0 = n$
8. $5 \times 7 = n$
9. $1 \times 6 = n$
10. $8 \times 6 = n$
11. $7 \times 7 = n$
12. $2 \times 9 = n$
13. $5 \times 8 = n$
14. $3 \times 4 = n$
15. $4 \times 9 = n$
16. $3 \times 3 = n$
17. $3 \times 0 = n$
18. $7 \times 9 = n$
19. $4 \times 5 = n$
20. $6 \times 6 = n$
21. $8 \times 4 = n$
22. $9 \times 6 = n$
23. $3 \times 1 = n$
24. $9 \times 9 = n$
25. $2 \times 6 = n$
26. $5 \times 9 = n$
27. $4 \times 4 = n$
28. $8 \times 0 = n$
29. $9 \times 5 = n$
30. $5 \times 5 = n$
31. $4 \times 0 = n$
32. $6 \times 8 = n$

Solve each story problem.

33. 9 boys need shoes. How many shoes?
34. 9 horses need shoes. How many shoes?
35. Apples: 7 cents each. 9 apples. How much money needed?
36. Each girl has 6 dolls. 7 girls. How many dolls?
37. Stamps: 8 cents each. 6 letters. How much money is needed?
38. 6 pencils in a set. 4 sets. How many pencils?
39. Each package of gum has 5 sticks. Bill has 7 packages. How many sticks of gum does Bill have?
40. A baseball team has 9 members. There are 8 teams playing. How many members are playing?
41. A triangle has 3 angles. Bobbie drew 7 triangles on her paper. How many angles did she form?

Reflected answers, Set 29: 1. 0, 17. 12, 18. 48, 10. 6, 9. 6, 2. 12, 5. 15, 1. 36, 34. 63, 33. 18, 32. 45, 31. 56, 30. 25, 29. 63, 18.

Set 30   *For use with page 161*

Solve the equations.

1. $9 \times 6 = n$
2. $5 \times n = 35$
3. $n \times 4 = 28$
4. $n \times 5 = 0$
5. $56 = 7 \times n$
6. $6 \times n = 24$
7. $n \times 8 = 32$
8. $9 \times n = 36$
9. $8 \times n = 8$
10. $n \times 3 = 21$

11. $4 \times n = 20$
12. $30 = 6 \times n$
13. $72 = 8 \times n$
14. $56 = n \times 8$
15. $n \times 9 = 63$
16. $7 \times n = 49$
17. $8 \times n = 40$
18. $7 \times 6 = n$
19. $30 = 5 \times n$
20. $n \times 7 = 28$

21. $n \times 5 = 25$
22. $21 = 7 \times n$
23. $24 = n \times 6$
24. $n \times 8 = 64$
25. $9 \times n = 72$
26. $n \times 9 = 54$
27. $48 = n \times 6$
28. $7 \times n = 42$
29. $35 = 7 \times n$
30. $8 \times n = 0$

31. $n \times 9 = 45$
32. $24 = 3 \times n$
33. $n \times 9 = 45$
34. $n \times 2 = 14$
35. $48 = 6 \times n$
36. $n \times 3 = 27$
37. $4 \times n = 24$
38. $32 = n \times 8$
39. $0 = 5 \times n$
40. $35 = 7 \times n$

Solve each story problem.

41. Candy bars are packed in bags of 6. Janet bought 8 bags. How many bars did she buy?

42. Children are seated in groups of 8. There are 9 groups in the library. How many children are in the library?

43. Pete colored 7 designs. Each design has 6 parts. How many parts did Pete color?

44. There are 4 quarts in each gallon. Mother bought 3 gallons of milk. How many quarts of milk did she buy?

45. Betty put pictures in a book. She put 5 on each page. She used 9 pages. How many pictures did she put in her book?

46. Bill has 8 nickels and 7 pennies. How much does he have?

Reflected answers, Set 30: 1. 54, 2. 7, 11. 5, 12. 5, 21. 5, 22. 3, 31. 5, 32. 8, 41. 48

Set 31   *For use with page 177*

Make sets of dots to help answer each of the following.
Then solve each equation.

1. How many sets of 4 are in a set of 12?
   $12 \div 4 = n$

2. How many sets of 3 are in a set of 6?
   $6 \div 3 = n$

3. How many sets of 5 are in a set of 30?
   $30 \div 5 = n$

4. How many sets of 7 are in a set of 21?
   $21 \div 7 = n$

Solve the equations. Use number lines to help.

5. $24 \div 6 = n$
6. $12 \div 2 = n$
7. $40 \div 5 = n$
8. $16 \div 4 = n$
9. $15 \div 3 = n$
10. $48 \div 8 = n$
11. $48 \div 6 = n$
12. $18 \div 3 = n$
13. $27 \div 9 = n$
14. $14 \div 7 = n$
15. $36 \div 9 = n$
16. $24 \div 8 = n$
17. $30 \div 10 = n$
18. $20 \div 2 = n$
19. $54 \div 9 = n$
20. $56 \div 7 = n$

Solve each story problem.

21. 24 children.
    4 in each group.
    How many groups?

22. 32 marbles.
    4 in each circle.
    How many circles?

23. 35 cents.
    5 cents for each boy.
    How many boys?

24. 28 dolls.
    7 in each box.
    How many boxes?

25. There are 48 cars in the parking lot. The cars are parked in rows of 8. How many rows in the lot?

Reflected answers, Set 31: 1. 3, 2. 2, 5. 4, 9. 5, 13. 3, 17. 3

A-22

Set 32 — For use with page 179

Find the quotients. Use multiplication to check your answers.

1. $8 \div 4 = n$
2. $12 \div 4 = n$
3. $36 \div 6 = n$
4. $27 \div 9 = n$
5. $7 \div 7 = n$
6. $30 \div 5 = n$
7. $24 \div 6 = n$
8. $12 \div 2 = n$
9. $27 \div 3 = n$
10. $56 \div 8 = n$
11. $0 \div 5 = n$
12. $9 \div 1 = n$
13. $64 \div 8 = n$
14. $18 \div 3 = n$
15. $36 \div 4 = n$
16. $10 \div 5 = n$
17. $54 \div 9 = n$
18. $32 \div 4 = n$
19. $28 \div 7 = n$
20. $24 \div 4 = n$
21. $16 \div 8 = n$
22. $14 \div 2 = n$
23. $10 \div 2 = n$
24. $54 \div 6 = n$
25. $63 \div 7 = n$
26. $56 \div 8 = n$
27. $81 \div 9 = n$
28. $9 \div 3 = n$
29. $0 \div 3 = n$
30. $36 \div 6 = n$
31. $45 \div 5 = n$
32. $24 \div 8 = n$
33. $12 \div 6 = n$
34. $4 \div 4 = n$
35. $35 \div 7 = n$
36. $16 \div 2 = n$
37. $49 \div 7 = n$
38. $42 \div 6 = n$
39. $64 \div 8 = n$
40. $15 \div 3 = n$

Reflected answers, Set 32: (inverted)

Set 33 — For use with page 185

Find the products.

1. $6 \times 5 = n$
2. $8 \times 3 = n$
3. $2 \times 7 = n$
4. $4 \times 6 = n$
5. $1 \times 8 = n$
6. $9 \times 7 = n$
7. $9 \times 8 = n$
8. $2 \times 1 = n$
9. $0 \times 6 = n$
10. $3 \times 3 = n$
11. $4 \times 8 = n$
12. $7 \times 6 = n$

Find the missing factors.

13. $5 \times n = 20$
14. $n \times 3 = 18$
15. $n \times 0 = 0$
16. $9 \times n = 72$
17. $4 \times n = 32$
18. $n \times 5 = 40$
19. $7 \times n = 7$
20. $n \times 3 = 21$
21. $n \times 9 = 54$
22. $n \times 8 = 56$
23. $1 \times n = 12$
24. $3 \times n = 12$

Find the quotients.

25. $16 \div 8 = n$
26. $12 \div 6 = n$
27. $72 \div 8 = n$
28. $21 \div 7 = n$
29. $0 \div 2 = n$
30. $4 \div 1 = n$
31. $63 \div 7 = n$
32. $48 \div 6 = n$
33. $24 \div 3 = n$
34. $8 \div 2 = n$
35. $14 \div 7 = n$
36. $16 \div 4 = n$
37. $81 \div 9 = n$
38. $18 \div 2 = n$
39. $6 \div 6 = n$
40. $64 \div 8 = n$

Reflected answers, Set 33: (inverted)

Set 34 For use with page 187

Find the missing factors and quotients.

1. $27 \div 3 = n$
2. $48 \div n = 6$
3. $n \div 4 = 4$
4. $72 \div 8 = n$
5. $n = 63 \div 7$
6. $48 \div 8 = n$
7. $32 \div n = 4$
8. $n \div 9 = 5$
9. $n \div 7 = 7$
10. $54 \div n = 6$
11. $7 = n \div 3$
12. $72 \div n = 8$
13. $3 = 21 \div n$
14. $81 \div n = 9$
15. $40 \div 8 = n$
16. $n \div 7 = 3$
17. $20 \div 4 = n$
18. $n = 42 \div 6$
19. $n \div 6 = 4$
20. $72 \div n = 9$
21. $n = 36 \div 4$
22. $n \div 3 = 8$
23. $30 \div n = 6$
24. $40 \div n = 5$
25. $5 = 15 \div n$
26. $n \div 8 = 0$
27. $42 \div 7 = n$
28. $9 \div n = 9$
29. $56 \div 8 = n$
30. $n = 24 \div 4$
31. $28 \div n = 4$
32. $n \div 3 = 4$
33. $48 \div n = 8$
34. $81 \div n = 9$
35. $n = 63 \div 9$
36. $n \div 6 = 9$
37. $72 \div n = 8$
38. $7 = 35 \div n$
39. $n = 12 \div 2$
40. $64 \div n = 8$

Solve each story problem.

41. 18 coins.
    9 in each set.
    How many sets?

42. 24 coins.
    8 sets.
    How many in each set?

43. 48 cookies.
    8 trays.
    How many on each tray?

44. 42 Girl Scouts.
    7 cars.
    How many in each car?

45. 72 books.
    8 for each boy.
    How many boys?

46. 32 cents.
    Milk: 4 cents.
    How many cartons?

47. Bill put his baseball cards into stacks of 4. Bill has 36 cards. How many stacks does he have?

Set 35 — For use with page 189

Find the quotients.

1. $2 \div 2$
2. $6 \div 3$
3. $16 \div 4$
4. $9 \div 1$
5. $14 \div 2$
6. $6 \div 2$
7. $2 \div 1$
8. $42 \div 6$
9. $63 \div 9$
10. $32 \div 4$
11. $24 \div 6$
12. $36 \div 6$
13. $8 \div 2$
14. $18 \div 2$
15. $6 \div 6$
16. $18 \div 3$
17. $27 \div 9$
18. $3 \div 3$
19. $64 \div 8$
20. $35 \div 7$
21. $12 \div 2$
22. $49 \div 7$
23. $48 \div 6$
24. $15 \div 5$
25. $72 \div 8$
26. $4 \div 1$
27. $24 \div 3$
28. $32 \div 8$
29. $30 \div 6$
30. $6 \div 1$
31. $56 \div 7$
32. $28 \div 4$
33. $8 \div 8$
34. $12 \div 4$
35. $7 \div 7$
36. $4 \div 2$
37. $9 \div 3$
38. $1 \div 1$
39. $21 \div 7$
40. $63 \div 7$
41. $12 \div 3$
42. $25 \div 5$
43. $20 \div 5$
44. $56 \div 8$
45. $16 \div 8$
46. $20 \div 4$
47. $15 \div 3$
48. $0 \div 1$
49. $18 \div 6$
50. $45 \div 9$
51. $10 \div 2$
52. $3 \div 1$
53. $48 \div 8$
54. $5 \div 5$
55. $7 \div 1$
56. $24 \div 8$
57. $42 \div 7$
58. $21 \div 3$
59. $30 \div 6$
60. $27 \div 3$
61. $24 \div 4$
62. $18 \div 9$
63. $28 \div 7$
64. $10 \div 5$
65. $9 \div 9$
66. $35 \div 5$
67. $54 \div 6$
68. $0 \div 9$
69. $5 \div 1$
70. $8 \div 1$
71. $4 \div 4$
72. $36 \div 4$
73. $40 \div 5$
74. $16 \div 2$
75. $0 \div 5$
76. $40 \div 8$
77. $36 \div 9$
78. $14 \div 7$
79. $12 \div 6$
80. $72 \div 8$
81. $72 \div 9$
82. $30 \div 5$
83. $81 \div 9$
84. $8 \div 4$
85. $48 \div 8$

Solve each story problem.

86. How many pennies equal 7 nickels?
87. How many weeks are 49 days?
88. Alice drew several triangles. She counted 27 angles. How many triangles did she draw?

Set 36   *For use with page 191*

Solve each equation.

1. $28 \div 4 = n$
2. $n = 54 \div 6$
3. $n \div 6 = 6$
4. $5 = 40 \div n$
5. $25 \div 5 = n$
6. $n = 72 \div 9$
7. $7 = 49 \div n$
8. $35 \div 5 = n$
9. $16 \div n = 2$
10. $32 \div n = 8$
11. $49 \div 7 = n$
12. $63 \div n = 9$
13. $54 \div n = 6$
14. $n = 7 \div 7$
15. $n \div 6 = 5$
16. $n \div 4 = 7$
17. $n = 36 \div 4$
18. $56 \div n = 7$
19. $n \div 3 = 4$
20. $27 \div n = 3$
21. $6 = n \div 6$
22. $48 \div 6 = n$
23. $n \div 8 = 1$
24. $28 \div n = 4$
25. $54 \div 9 = n$
26. $40 \div 8 = n$
27. $n = 30 \div 5$
28. $48 \div n = 6$
29. $n \div 8 = 5$
30. $16 \div n = 4$
31. $n \div 2 = 10$
32. $42 \div n = 7$
33. $63 \div n = 7$
34. $15 \div n = 5$
35. $n \div 2 = 4$
36. $n \div 3 = 9$

Solve each story problem.

37. 42 dots.
    6 sets.
    How many in each set?

38. 24 dots.
    6 in each set.
    How many sets?

39. 27 feet.
    3 feet per yard.
    How many yards?

40. 16 pints.
    2 pints per quart.
    How many quarts?

41. Passengers flying in a 707 airplane sit in rows of 6. 48 passengers can sit in the "no smoking" section. How many rows have "no smoking"?

42. The book store is having an 8¢ sale on 10¢ tablets. Shelly has 56¢. How many tablets can she buy?

43. How many 6-bottle cartons are needed to pack 54 bottles?

Reflected answers, Set 36: 1. 7, 2. 9, 10. 4, 11. 7, 19. 12, 21. 36, 27. 6, 28. 8, 29. 40, 37. 7, 38. 4

Set 37   *For use with page 223*

Tell whether the sentence is true (T) or false (F).

1. 16 is a multiple of 8.
2. 9 is a factor of 18.
3. 4 is a factor of 6.
4. 3 is a factor of 12.
5. 7 is a prime number.
6. 12 is a multiple of 6.
7. 10 is a multiple of 3.
8. 15 is a prime number.
9. 0 is a multiple of 6.
10. 0 is a factor of 6.
11. 35 is a prime number.
12. 54 is a multiple of 8.
13. 9 is a factor of both 36 and 72.
14. 42 is a multiple of both 7 and 5.
15. A prime number has more than two factors.
16. 18 has exactly 6 factors.
17. 24 is a multiple of 3, 6, and 8.
18. 40 is a multiple of 0, 10, 20, and 40.

Complete the sentence.

19. Since 5 × 8 = 40, 5 and 8 are _____ of 40.
20. Since 6 × 9 = 54, 54 is a multiple of 6 and _____.
21. The first 5 multiples of 9 are _____.
22. The factors of 4 and 8 are _____.
23. The first 3 multiples of both 2 and 3 are _____.
24. The factors of 37 are _____.
25. The multiples of 6 between 20 and 40 are _____.
26. The prime numbers between 20 and 30 are _____.
27. The factors of 36 that are between 10 and 20 are _____.

Reflected answers, Set 37: 1. T, 2. T, 5. T, 7. F, 8. F, 19. factors, 20. 9

Set 38 *For use with page 229*

Find the value in cents.

1. 3 dimes
2. 7 dimes
3. 12 dimes
4. 18 dimes
5. 25 dimes
6. 11 dimes
7. 32 dimes
8. 85 dimes
9. 91 dimes
10. 1 dime
11. 75 dimes
12. 0 dimes
13. 67 dimes
14. 51 dimes
15. 87 dimes

Solve the equations.

16. $(20 \times 10) + (3 \times 10) = n$
17. $(60 \times 10) + (4 \times 10) = n$
18. $(30 \times 10) + (7 \times 10) = n$
19. $(80 \times 10) + (1 \times 10) = n$
20. $(70 \times 10) + (5 \times 10) = n$
21. $(40 \times 10) + (6 \times 10) = n$
22. $(10 \times 10) + (9 \times 10) = n$
23. $(90 \times 10) + (2 \times 10) = n$

24. $18 \times 10 = n$
25. $24 \times 10 = n$
26. $37 \times 10 = n$
27. $7 \times 10 = n$
28. $97 \times 10 = n$
29. $48 \times 10 = n$
30. $21 \times 10 = n$
31. $77 \times 10 = n$
32. $31 \times 10 = n$
33. $10 \times 11 = n$
34. $10 \times 89 = n$
35. $10 \times 63 = n$
36. $10 \times n = 90$
37. $n \times 10 = 40$
38. $10 \times n = 120$
39. $50 \times n = 500$
40. $n \times 22 = 220$
41. $36 \times n = 360$
42. $10 \times n = 270$
43. $n \times 10 = 570$
44. $n \times 10 = 970$

Find the products.

45. 9×10
46. 7×10
47. 10×8
48. 10×0
49. 2×10
50. 10×6
51. 1×10
52. 14×10
53. 10×16
54. 8×10
55. 10×7
56. 6×10
57. 10×4
58. 0×10
59. 3×10
60. 12×10
61. 15×10
62. 10×18
63. 34×10
64. 52×10
65. 65×10
66. 20×10
67. 53×10
68. 10×81
69. 40×10
70. 10×80
71. 38×100
72. 65×100
73. 100×73
74. 32×10
75. 10×58
76. 36×10
77. 58×10
78. 76×10
79. 10×40
80. 62×10

A-28

Set 39   For use with page 231

Find the products.

1. 3 × 40	13. 80 × 6	25. 3 × 20	37. 90 × 5
2. 60 × 4	14. 60 × 6	26. 7 × 60	38. 60 × 9
3. 50 × 9	15. 8 × 60	27. 30 × 9	39. 70 × 5
4. 8 × 70	16. 7 × 70	28. 6 × 30	40. 90 × 9
5. 2 × 30	17. 80 × 8	29. 5 × 80	41. 275 × 10
6. 8 × 50	18. 90 × 6	30. 6 × 60	42. 10 × 832
7. 20 × 9	19. 8 × 20	31. 8 × 90	43. 653 × 10
8. 7 × 30	20. 90 × 7	32. 5 × 60	44. 10 × 976
9. 4 × 80	21. 4 × 50	33. 70 × 8	45. 8345 × 10
10. 50 × 6	22. 70 × 3	34. 80 × 7	46. 8 × 100
11. 9 × 90	23. 60 × 8	35. 4 × 90	47. 100 × 9
12. 4 × 70	24. 7 × 90	36. 4 × 50	48. 18 × 100

Reflected answers, Set 39: 1. 120, 2. 240, 13. 480, 14. 360, 25. 60, 26. 420, 37. 450, 38. 540.

Set 40   For use with page 239

Solve the equations.

1. 5 × 18 = (5 × 10) + (5 × *n*)
2. 7 × 16 = (7 × *n*) + (7 × 6)
3. 4 × 22 = (4 × 20) + (4 × *n*)
4. 8 × 27 = (8 × *n*) + (8 × 7)
5. 6 × 35 = (6 × 30) + (6 × *n*)
6. 9 × 42 = (9 × *n*) + (9 × 2)
7. 7 × 53 = (7 × 50) + (7 × *n*)
8. 2 × 83 = (2 × *n*) + (2 × 3)
9. 3 × 75 = (3 × 70) + (3 × *n*)
10. 9 × 62 = (9 × *n*) + (9 × 2)
11. (5 × 10) + (5 × 3) = *n*
12. (6 × 10) + (6 × 8) = *n*
13. (4 × 20) + (4 × 2) = *n*
14. (5 × 20) + (5 × 9) = *n*
15. (3 × 30) + (3 × 8) = *n*
16. (8 × 30) + (8 × 2) = *n*
17. (7 × 60) + (7 × 1) = *n*
18. (6 × 50) + (6 × 7) = *n*
19. (4 × 70) + (4 × 6) = *n*
20. (9 × 40) + (9 × 7) = *n*
21. (2 × 90) + (2 × 7) = *n*
22. (7 × 80) + (7 × 8) = *n*
23. (8 × 60) + (8 × 7) = *n*
24. (9 × 90) + (9 × 3) = *n*

Reflected answers, Set 40: 1. 8, 2. 10, 3. 2, 13. 88, 14. 145, 15. 114.

Set 41   *For use with page 241*

Find the products.

1. 32 ×2	2. 16 ×4	3. 45 ×5	4. 41 ×4	5. 28 ×5	6. 37 ×3
7. 27 ×4	8. 81 ×5	9. 90 ×3	10. 65 ×8	11. 56 ×2	12. 43 ×9
13. 84 ×4	14. 52 ×7	15. 63 ×8	16. 98 ×3	17. 21 ×9	18. 79 ×2
19. 11 ×3	20. 22 ×5	21. 53 ×4	22. 43 ×6	23. 51 ×3	24. 24 ×7
25. 45 ×2	26. 67 ×7	27. 81 ×6	28. 60 ×5	29. 79 ×8	30. 85 ×9

Solve each story problem.

31. Mr. Williams is planting pear trees. He plants 16 rows. Each row has 8 trees. How many trees did he plant?

32. Model airplane kits have 24 pieces in them. Mario buys 7 kits. How many airplane pieces does Mario have?

33. A building has 38 floors. Each floor has 9 offices. How many offices are in the building?

34. Candy is 79 cents a pound. How much will Nancy have to pay for 6 pounds of candy?

35. Mike collects coins. Each card contains 64 coins. Mike has 5 cards filled. How many coins does he have?

Set 42 For use with page 243

Find the products.

1. 19 ×2	2. 62 ×4	3. 35 ×3	4. 13 ×4	5. 16 ×2	6. 22 ×4
7. 29 ×6	8. 43 ×7	9. 78 ×3	10. 13 ×3	11. 32 ×5	12. 34 ×4
13. 13 ×7	14. 24 ×5	15. 32 ×4	16. 63 ×8	17. 99 ×9	18. 54 ×3
19. 63 ×8	20. 76 ×6	21. 67 ×4	22. 46 ×2	23. 53 ×9	24. 77 ×8
25. 98 ×7	26. 39 ×2	27. 55 ×6	28. 80 ×5	29. 39 ×3	30. 79 ×9

Solve each story problem.

31. Nancy delivers 78 papers a day. How many papers does she deliver in 7 days?

32. There are 24 hours in one day. How many hours are in 6 days?

33. There are 48 jars of baby food in a case. Mrs. Fong bought 9 cases. How many jars did she buy?

34. A school bus can carry 68 children. How many children can ride in 8 buses this size?

35. During his vacation trip, Mr. Raymond drove his car an average of 57 miles per hour. He drove for 4 hours on Friday and 5 hours on Saturday. How far did Mr. Raymond drive?

Set 43 *For use with page 247*

Find each product.

1. 47 ×2	2. 56 ×3	3. 75 ×5	4. 86 ×4	5. 93 ×4	6. 76 ×3
7. 88 ×7	8. 96 ×8	9. 47 ×6	10. 52 ×9	11. 61 ×8	12. 78 ×7
13. 42 ×5	14. 65 ×8	15. 82 ×3	16. 97 ×2	17. 94 ×7	18. 58 ×4
19. 74 ×6	20. 83 ×4	21. 86 ×9	22. 71 ×6	23. 88 ×8	24. 49 ×7
25. 38 ×8	26. 92 ×6	27. 79 ×9	28. 97 ×2	29. 65 ×4	30. 87 ×5

Solve each story problem.

31. The assembly hall has 78 rows of seats. There are 9 seats in each row. How many persons can sit in the hall?

32. There are 8 rows of bricks in a wall. Each row contains 98 bricks. How many bricks are in the wall?

33. Paper costs 89 cents a tablet. Pens cost 29 cents each. How much do 7 tablets and 4 pens cost?

34. The basketball team practices 85 minutes on Monday and Wednesday. They practice 75 minutes on Tuesday, Thursday and Friday. How many minutes does the team practice each week?

35. Mike builds bookcases with 6 shelves. Each shelf holds 16 books. How many books can be put in 8 bookcases?

Set 44 — For use with page 249

Find the products.

1. 413 ×3	2. 525 ×2	3. 334 ×4	4. 242 ×5	5. 453 ×3	6. 264 ×2
7. 563 ×9	8. 652 ×3	9. 835 ×8	10. 496 ×7	11. 741 ×6	12. 416 ×4
13. 123 ×2	14. 564 ×3	15. 345 ×4	16. 153 ×5	17. 254 ×6	18. 347 ×4
19. 243 ×3	20. 563 ×5	21. 674 ×7	22. 755 ×2	23. 486 ×9	24. 327 ×8
25. 1542 ×2		26. 3361 ×3		27. 5782 ×5	28. 7954 ×3
29. 1738 ×3		30. 2138 ×5		31. 3279 ×6	32. 4960 ×9
33. 4189 ×5		34. 7498 ×6		35. 5276 ×7	36. 1556 ×8

Solve each story problem.

37. A certain truck company allows their truckers to drive 375 miles per day. How far can one trucker drive in 6 days?

38. The average weight of each lineman on a certain professional football team is 238 pounds. There are 7 men in the line. How much does the line weigh?

39. The 9 children in the Tiger Club collected paper for a paper drive. Each child collected 108 pounds of paper per week. The drive lasted 6 weeks. How many pounds of paper was collected by the Tiger Club?

Reflected answers, Set 44: 1. 1239, 2. 1050, 3. 1336, 4. 1210, 5. 1359, 6. 528, 25. 3084, 26. 10,083, 27. 28,910, 28. 23,862

Set 45 *For use with page 279*

Find the products.

1. 7 × 9	5. 7 × 4	9. 3 × 40	13. 90 × 2	17. 200 × 3
2. 8 × 6	6. 20 × 6	10. 7 × 50	14. 30 × 7	18. 100 × 4
3. 3 × 5	7. 40 × 8	11. 8 × 10	15. 70 × 6	19. 700 × 6
4. 2 × 9	8. 7 × 70	12. 9 × 80	16. 800 × 7	20. 8 × 900

Find the quotients.

21. 72 ÷ 8	25. 30 ÷ 5	29. 180 ÷ 2	33. 300 ÷ 6	37. 5400 ÷ 9
22. 64 ÷ 8	26. 560 ÷ 7	30. 210 ÷ 7	34. 400 ÷ 5	38. 1000 ÷ 5
23. 63 ÷ 7	27. 270 ÷ 3	31. 480 ÷ 6	35. 60 ÷ 3	39. 2800 ÷ 7
24. 24 ÷ 6	28. 180 ÷ 3	32. 360 ÷ 4	36. 2100 ÷ 3	40. 6300 ÷ 9

Find the missing factors.

41. 5 × n = 35	46. 8 × n = 320	51. 2 × n = 1000	56. n × 5 = 1500
42. n × 2 = 16	47. n × 5 = 450	52. n × 6 = 5400	57. n × 7 = 4200
43. 6 × n = 18	48. 7 × n = 70	53. 9 × n = 6300	58. 6 × n = 4200
44. n × 9 = 36	49. n × 8 = 640	54. n × 4 = 2800	59. n × 2 = 1400
45. n × 7 = 280	50. 3 × n = 150	55. n × 9 = 3600	60. 9 × n = 7200

Find the missing quotients.

61. 48 ÷ 6 = n	66. 270 ÷ 3 = n	71. 60 ÷ 2 = n	76. 3200 ÷ 4 = n
62. 20 ÷ 4 = n	67. 210 ÷ 7 = n	72. 120 ÷ 3 = n	77. 1500 ÷ 5 = n
63. 16 ÷ 8 = n	68. 360 ÷ 9 = n	73. 4800 ÷ 6 = n	78. 2400 ÷ 3 = n
64. 56 ÷ 7 = n	69. 300 ÷ 5 = n	74. 4900 ÷ 7 = n	79. 5600 ÷ 8 = n
65. 120 ÷ 2 = n	70. 320 ÷ 8 = n	75. 4500 ÷ 9 = n	80. 8100 ÷ 9 = n

Reflected answers, Set 45: 1. 63, 5. 28, 9. 120, 13. 180, 17. 600, 21. 9, 25. 6, 29. 90, 33. 50, 37. 600, 41. 7, 45. 40, 51. 500, 55. 300, 61. 8, 65. 60, 71. 30, 75. 500

Set 46 — For use with page 289

Find the quotients.

1. 36 ÷ 2	10. 168 ÷ 7	19. 117 ÷ 3	28. 296 ÷ 8	37. 336 ÷ 8
2. 48 ÷ 4	11. 39 ÷ 3	20. 162 ÷ 6	29. 153 ÷ 3	38. 279 ÷ 9
3. 96 ÷ 8	12. 74 ÷ 2	21. 272 ÷ 8	30. 232 ÷ 4	39. 364 ÷ 7
4. 72 ÷ 6	13. 189 ÷ 9	22. 140 ÷ 5	31. 265 ÷ 5	40. 504 ÷ 7
5. 65 ÷ 5	14. 108 ÷ 4	23. 210 ÷ 5	32. 180 ÷ 4	41. 336 ÷ 4
6. 77 ÷ 7	15. 174 ÷ 6	24. 287 ÷ 7	33. 252 ÷ 6	42. 456 ÷ 6
7. 52 ÷ 4	16. 175 ÷ 7	25. 387 ÷ 9	34. 203 ÷ 7	43. 364 ÷ 4
8. 88 ÷ 4	17. 258 ÷ 6	26. 198 ÷ 9	35. 333 ÷ 9	44. 376 ÷ 8
9. 110 ÷ 5	18. 132 ÷ 4	27. 368 ÷ 8	36. 112 ÷ 2	45. 469 ÷ 7

Reflected answers, Set 46: 1. 18, 2. 12, 3. 12, 10. 24, 11. 13, 12. 37, 19. 39, 20. 27, 21. 34, 28. 37, 29. 51, 30. 58, 37. 42, 38. 31, 39. 52

Set 47 — For use with page 291

Solve each story problem.

1. 112 players. 8 on each team. How many teams?
2. 252 books. 6 shelves. How many on each shelf?
3. 85 cents. All nickels. How many nickels?
4. 364 children. 7 per car. How many cars?
5. 184 truck tires sold. 4 tires for each truck. How many trucks?
6. 243 balloons. 9 packs. How many in each pack?
7. 228 bottles. 6 per carton. How many cartons?
8. 336 seats. 8 rows. How many in each row?
9. 336 days. How many weeks?
10. 639 pounds of shrimp. 9 pounds per carton. How many cartons?

Reflected answers, Set 47: 1. 14, 2. 42, 6. 27, 7. 38

Set 48 For use with page 295

Find the quotients.

1. 2)$\overline{24}$	2. 3)$\overline{63}$	3. 4)$\overline{48}$	4. 6)$\overline{96}$	5. 7)$\overline{84}$	6. 5)$\overline{80}$
7. 3)$\overline{78}$	8. 5)$\overline{75}$	9. 7)$\overline{98}$	10. 4)$\overline{144}$	11. 2)$\overline{88}$	12. 3)$\overline{138}$
13. 4)$\overline{52}$	14. 2)$\overline{32}$	15. 3)$\overline{48}$	16. 5)$\overline{60}$	17. 6)$\overline{72}$	18. 7)$\overline{91}$
19. 7)$\overline{112}$	20. 8)$\overline{176}$	21. 9)$\overline{207}$	22. 6)$\overline{78}$	23. 8)$\overline{96}$	24. 9)$\overline{108}$
25. 2)$\overline{108}$	26. 4)$\overline{92}$	27. 6)$\overline{204}$	28. 3)$\overline{171}$	29. 4)$\overline{272}$	30. 5)$\overline{235}$
31. 5)$\overline{470}$	32. 3)$\overline{222}$	33. 2)$\overline{134}$	34. 7)$\overline{385}$	35. 8)$\overline{752}$	36. 6)$\overline{588}$
37. 7)$\overline{301}$	38. 9)$\overline{585}$	39. 5)$\overline{415}$	40. 6)$\overline{564}$	41. 5)$\overline{445}$	42. 4)$\overline{308}$
43. 8)$\overline{368}$	44. 6)$\overline{342}$	45. 4)$\overline{384}$	46. 7)$\overline{483}$	47. 8)$\overline{312}$	48. 9)$\overline{666}$

Reflected answers, Set 48: 1. 12, 2. 21, 3. 12, 4. 16, 5. 12, 6. 16, 25. 54, 26. 23, 27. 34, 28. 57, 29. 68, 30. 47

Set 49 For use with page 301

Find the quotients and remainders.

1. 2)$\overline{57}$	2. 3)$\overline{46}$	3. 5)$\overline{82}$	4. 4)$\overline{77}$	5. 7)$\overline{98}$	6. 6)$\overline{86}$
7. 2)$\overline{69}$	8. 3)$\overline{92}$	9. 4)$\overline{58}$	10. 5)$\overline{76}$	11. 6)$\overline{69}$	12. 7)$\overline{87}$
13. 5)$\overline{235}$	14. 2)$\overline{146}$	15. 5)$\overline{478}$	16. 6)$\overline{379}$	17. 7)$\overline{598}$	18. 4)$\overline{192}$
19. 3)$\overline{163}$	20. 6)$\overline{475}$	21. 3)$\overline{286}$	22. 4)$\overline{379}$	23. 4)$\overline{281}$	24. 2)$\overline{192}$
25. 9)$\overline{654}$	26. 8)$\overline{355}$	27. 7)$\overline{463}$	28. 3)$\overline{264}$	29. 8)$\overline{789}$	30. 9)$\overline{237}$
31. 5)$\overline{142}$	32. 2)$\overline{173}$	33. 3)$\overline{274}$	34. 8)$\overline{274}$	35. 5)$\overline{385}$	36. 7)$\overline{496}$
37. 8)$\overline{572}$	38. 7)$\overline{351}$	39. 2)$\overline{138}$	40. 3)$\overline{249}$	41. 4)$\overline{354}$	42. 9)$\overline{665}$
43. 2)$\overline{187}$	44. 6)$\overline{496}$	45. 6)$\overline{576}$	46. 8)$\overline{791}$	47. 4)$\overline{284}$	48. 9)$\overline{823}$

Reflected answers, Set 49: 1. 28 R1, 2. 15 R1, 3. 16 R2, 4. 19 R1, 5. 14 R0, 6. 14 R2, 25. 72 R6, 26. 44 R3, 27. 66 R1, 28. 88 R0, 29. 98 R5, 30. 26 R3

A-36

Set 50 For use with page 303

Find the quotients.

1. 102 ÷ 6	5. 243 ÷ 9	9. 273 ÷ 3	13. 396 ÷ 4	17. 260 ÷ 5
2. 92 ÷ 4	6. 140 ÷ 5	10. 301 ÷ 7	14. 434 ÷ 7	18. 837 ÷ 9
3. 195 ÷ 3	7. 217 ÷ 7	11. 423 ÷ 9	15. 608 ÷ 8	19. 784 ÷ 8
4. 256 ÷ 8	8. 384 ÷ 8	12. 402 ÷ 6	16. 170 ÷ 2	20. 320 ÷ 5

Find the quotients and remainders.

21. 77 ÷ 3	24. 91 ÷ 2	27. 270 ÷ 4	30. 715 ÷ 9	33. 699 ÷ 9
22. 87 ÷ 6	25. 275 ÷ 8	28. 428 ÷ 8	31. 517 ÷ 7	34. 775 ÷ 8
23. 190 ÷ 7	26. 192 ÷ 5	29. 259 ÷ 3	32. 538 ÷ 6	35. 438 ÷ 5

Find the sums, differences, products, and quotients.

36. 654 −276	37. 9)477	38. 364 ×7	39. 7)448	40. 719 −263
41. 923 −357	42. 8)596	43. 539 +885	44. 9)396	45. 396 ×6
46. 466 ×8	47. 6)564	48. 927 +878	49. 4)373	50. 832 −578
51. 394 +934	52. 8)440	53. 971 −796	54. 3)261	55. 356 ×8

Solve each story problem.

56. Nancy has 185 stamps. She places 8 on each page of her book. How many pages are full? How many more stamps are needed to fill the next page?

Reflected answers, Set 50: 1. 17, 5. 27, 9. 91, 13. 99, 17. 52, 21. 25 R2, 24. 45 R1, 27. 67 R2, 30. 79 R4, 33. 77 R6, 36. 378, 37. 53, 38. 2548, 39. 64, 40. 456

A-37

Books to Explore

Adler, Irving. *The Giant Golden Book of Mathematics.*
   New York, Golden Press, 1960.
Have you ever wondered how a tree grows or why a volcano is shaped as it is or what makes a card trick work? This colorful book answers these and many other questions, through exploring the world of mathematics. You'll find all kinds of exciting ideas about numbers and what they mean in our daily lives. Here are just a few of the interesting topics:

How the Mayan Indians wrote 100	13
The puzzle of the King's reward	21
Bridges, planets and whispering galleries	54
Why a navigator needs a clock	62
A magical card trick	81

Brindze, Ruth. *The Story of Our Calendar.*
   New York, Vanguard, 1949.
Men did not always have calendars to tell what day of the month it was. In fact, they did not always know about months. It took thousands of years to develop the calendar as we know it, and each version presented new problems. This book tells you:

The difference between sun years and moon years	8
Who decreed the first leap year	36
How the word "calendar" was invented	40
The year the calendar was set back 11 whole days	47

Carona, Philip. *Things That Measure.*
   Englewood Cliffs, New Jersey, Prentice-Hall, Inc., 1962.
Many pictures and diagrams help the author tell you about different ways to measure things. He covers a wide range of subjects, including:

A unit called the cubit	13
Using tobacco for money	15
Who invented the first watch	47
Weather instruments	50-55

Smith, George O. *Mathematics: The Language of Science.*
   New York, G. P. Putnam's Sons, 1961.
Interesting stories show how mathematical language is a handy tool for scientists. You will learn how the first men kept track of their possessions, how Roman businessmen struggled with Roman numerals, how the Babylonians discovered place value, and how Copernicus had trouble convincing people that the sun was the center of the universe.
   Other stories you might enjoy reading in this book are:

The discovery of nothing	23
Square numbers	31
Mr. Newton's apple	47-49
Investigating a Mobius strip	63-64

Some shorter books to look for in your library are listed below.

Adler, Irving and Ruth. *Numbers Old and New*.
New York, John Day, 1960.
This book explains the counting methods of Australian natives and Mayan Indians, and the fractions used by Egyptians and Greeks. For fun, read the chapters about lucky and unlucky numbers, number tricks, and numbers in nature.

Bendick, Jeanne. *First Book of Time*.
New York, Watts, 1963.
Excellent pictures help tell the story of time and how to measure it. All kinds of clocks are described—from sun dials and water clocks to the atomic clock and clocks in your body.

Bendick, Jeanne, and **Levin, Marcia O.** *Take a Number*.
New York, Whittlesey House, 1961.
Stories and puzzles explain different ways of counting with numbers. Some are very easy, like counting on your fingers; others are more complicated, like using a computer.

Charosh, Mannis. *Straight Lines, Parallel Lines, Perpendicular Lines*.
New York, Thomas Y. Crowell, 1970.
By using a piece of string and a checkerboard, you can explore the world of straight, parallel and perpendicular lines.

Clarke, Mollie. *Beads. A Group of Children. Numbers. Dominoes. What is Missing? 20 Sticks. Houses. The Calendar. The Piggy Bank. A Box of Crayons. Square Inches. Sweets. Cakes and Candles. What Is Inside? Shapes. Buttons. A Dozen Eggs. Symmetrical Shapes*.
Newton, Massachusetts, Selective Educational Equipment, Inc.
There should be something here to interest you.

Friskey, Margaret. *The Mystery of the Farmer's Three Fives*.
Chicago, Children's Press, 1963.
A book about animals that compares groups in the barnyard. You'll enjoy the mystery.

Hine, Al. *Money Round the World*.
New York, Harcourt, Brace & World, Inc., 1963.
All about trading and money, from stones to metal coins and paper money.

Hoban, Tana. *Shapes and Things*.
New York, Macmillan Co., 1970.
The author shows simple, everyday objects in photograms—pictures made without a camera. This is a real treat for your eyes.

Hutchins, Pat. *Clocks and More Clocks*.
New York, Macmillan, 1970.
Have you ever tried to check the time on three different clocks and found three different answers? Look through this picture book for fun with clocks.

Jacobs, Leland B. *Delight in Number*.
New York, Holt, Rinehart and Winston, Inc., 1964.
These happy poems about groups of objects use many number words.

Kettlekamp, Larry. *Spirals*.
Englewood Cliffs, New Jersey, Prentice-Hall, Inc., 1964.
An enjoyable look at spirals—some are in nature, some man-made, and they go in and out, up and down.

Linn, Charles. *Estimation*.
New York, Thomas Y. Crowell, 1970.
Interesting experiments and activities help you improve your estimating skills.

Massoglia, Elinor. *Fun-Time Paper Folding*.
Chicago, Children's Press, 1959.
You would be surprised how many different things you can make by folding a single piece of paper. This book gives directions, and there is no cutting or pasting.

Rhodes, Dorothy. *How To Read a City Map*.
Chicago, Elk Grove Press, Inc., 1967.
Through pictures you can learn to read a map. How do you locate rivers, railroads, airports, historical landmarks, hospitals, roads, and government buildings on a map? How do you measure distances with a scale of miles? You can learn all this and more.

Russell, Solveig P. *One, Two, Three and Many: A First Look At Numbers*.
New York, Walck, 1970.
This book tells you all about the history of numbers and counting.

Sitomer, Mindel and Harry. *What Is Symmetry?*
New York, Thomas Y. Crowell, 1970.
With the help of the alligators, you can locate symmetries in nature and in man-made objects.

Savastava, Jane. *Weighing and Balancing*.
New York, Thomas Y. Crowell, 1970.
This book shows you how to make a simple balance.

Whitney, David C. *The Easy Book of Multiplication*.
New York, Franklin Watts, Inc., 1969.
This book will help you understand the process of multiplication with examples to show the facts.

Glossary

addend Any one of a set of numbers to be added. In the equation 4 + 5 = 9, the numbers 4 and 5 are addends.

addition An operation that combines a first number and a second number to give exactly one number. The two numbers are called addends, and the one number which is the result of combining the two numbers is called the sum of the addends.

angle Two rays from a single point.

approximation One number is an approximation of another number if the first number is suitably "close" (according to context) to the other number.

area The area of a closed figure or region is the measure of that region as compared to a given selected region called the unit, usually a square region in the case of area.

borrow A commonly used term for the regrouping process involved in certain types of subtraction.

Example:

$$\begin{array}{r}\overset{3}{\cancel{4}}\overset{13}{\cancel{3}}\\-1\;7\\\hline\end{array} \rightarrow \begin{array}{r}3\;0\\-1\;0\\\hline 2\;0\end{array} \quad \begin{array}{r}1\;3\\-\;\;\;7\\\hline 6\end{array} \rightarrow \begin{array}{r}4\;3\\-1\;7\\\hline 2\;6\end{array}$$

carry A commonly used term for the regrouping that is involved in addition.

Example:
$$\begin{array}{r}57\\+26\\\hline 83\end{array} \rightarrow \begin{array}{r}50\;+\;7\\20\;+\;6\\\hline 70\;+\;13\;=\;83\end{array}$$

centimeter A unit of length. One centimeter is $\frac{1}{100}$ meter.

chord A line segment that has its end points on a given circle.

circle A set of all points in a plane which are a specified distance from a given point called the center or center point.

compass A device for drawing models of a circle.

coordinate Number pair used in graphing.

coordinate axes Two number lines intersecting at right angles at 0.

count To name numbers in regular succession.

cube A rectangular prism (box) such that all faces are squares.

diagonal A segment joining two nonadjacent vertices of a polygon. In the figure, the diagonal is segment AB.

diameter A chord that passes through the center point of the circle.

difference The number resulting from the subtraction operation.

digits The basic Hindu-Arabic symbols used to write numerals. In the base-ten system, these are the digits 0, 1, 2, 3, 4, 5, 6, 7, 8, 9.

division An operation related to multiplication as illustrated:

$$3 \times 4 = 12 \begin{array}{l}\rightarrow 12 \div 3 = 4\\ \rightarrow 12 \div 4 = 3\end{array}$$

divisor In the problem 33 ÷ 7, 7 is called the divisor.

Example:
$$\text{divisor} \rightarrow \begin{array}{r}4\\7\overline{)33}\\28\\\hline 5\end{array}$$

empty set A set that has no objects in it.

equality (equals; or =) A mathematical relation of being exactly the same.

equation A mathematical sentence involving the use of the equality symbol. Examples: 5 + 4 = 9; 7 + □ = 8; n + 3 = 7.

equivalent fractions Two fractions are equivalent when it can be shown that they each can be used to represent the same amount of a given object. Also, two fractions are equivalent if these two products are the same:

$$\frac{3}{4} \underset{\longrightarrow 8}{\overset{\longrightarrow 6}{\times}} \begin{array}{l}\rightarrow 4 \times 6 \rightarrow 24\\ \rightarrow 3 \times 8 \rightarrow 24\end{array}$$

A-41

equivalent sets Two sets that may be placed in a one-to-one correspondence.

estimate To find an approximation for a given number. (Sometimes a sum, a product, etc.)

even numbers The whole-number multiples of 2 (0, 2, 4, 6, 8, 10, 12, . . .).

factor See multiplication. The equation $6 \times 7 = 42$ illustrates that both 6 and 7 are factors of 42.

fraction A symbol for a fractional number, usually written $\frac{2}{3}, \frac{3}{4}, \frac{1}{2}$, and so on.

graph (1) A set of points associated with a given set of numbers or set of number pairs. (2) A picture used to illustrate a given collection of data. The data might be pictured in the form of a bar graph, a circle graph, a line graph, or a pictograph. (3) To draw the graph of.

greater than ($>$) One of the two basic inequality relations. Examples: $8 > 5$, $28 > 25$, $80 > 50$.

grouping principle (associative principle) When adding (or multiplying) three numbers, you can change the grouping and the sum (or product) is the same.

Examples: $2 + (8 + 6) = (2 + 8) + 6$
$3 \times (4 \times 2) = (3 \times 4) \times 2$

hexagon A six-sided polygon.

hypotenuse The side opposite the right angle in a right triangle.

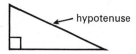

inch A unit of length. One inch is $\frac{1}{12}$ of a foot.

inequality ($<, \neq, >$) In arithmetic, a relation indicating that two numbers are not the same.

legs of a right triangle The two sides of a right triangle other than the hypotenuse.

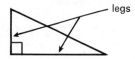

length (1) A number indicating the measure of one segment with respect to another segment, called the unit. (2) Sometimes used to denote one dimension (usually the greater) of a rectangle.

less than ($<$) One of the two basic inequality relations. Examples: $5 < 8$, $25 < 28$, $50 < 80$.

line A line is a set of points that "goes on and on" in both directions. There is only one line through any two points.

line segment See segment.

matching lines Lines used to indicate the correspondence between the objects in two sets.

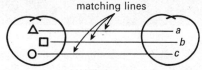

measure (1) A number indicating the relation between a given object and a suitable unit. (2) The process of finding the number described in (1).

midpoint A point that divides a line segment into two parts of the same size.

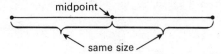

minus ($-$) Used to indicate the subtraction operation, as in $7 - 3 = 4$ (read, "7 minus 3 equals 4").

multiple A first number is a multiple of a second number if there is a whole number that multiplies by the second number to give the first number. Example: 24 is a multiple of 6 since $4 \times 6 = 24$.

multiplication An operation that combines a first number and a second number to give exactly one number. The two numbers are called factors, and the one number which is a result of combining the two numbers is called the product of the two numbers.

multiplication-addition principle (distributive principle) This principle is sometimes described in terms of "breaking apart" a number before multiplying. Example: $6 \times (20 + 4) = (6 \times 20) + (6 \times 4)$

negative number If a number adds to a whole number to give 0, it is a negative number.

For example: $5 + {}^-5 = 0$
$19 + {}^-19 = 0$

number line A line on which specified points are given number labels or names. The following example illustrates the whole-number line.

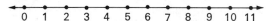

number pair Any pair of numbers. In this book, usually a pair of whole numbers.

numeral A symbol for a number.

odd number Any whole number that is not even.

one principle (for multiplication) Any number multiplied by 1 is that same number.

one-to-one correspondence A one-to-one correspondence exists between two sets when the elements of one can be matched with the elements of the other in such a way that each element of the first set is matched with exactly one element of the second set and each element of the second set is matched with exactly one element of the first set.

order principle (commutative principle) When adding (or multiplying) two numbers, the order of the addends (or factors) does not affect the sum (or product). Examples: $4 + 5 = 5 + 4$, $2 \times 3 = 3 \times 2$.

parallel lines Two lines which lie in the same plane and do not intersect.

parallelogram A quadrilateral with its opposite sides parallel.

parentheses A pair of curved symbols, (), used to indicate grouping or order of performing operations. Examples:

$(5 \times 4) - 2 = 18; \quad 5 \times (4 - 2) = 10.$

pentagon A five-sided polygon.

place value A system used for writing numerals for numbers, using only a definite number of symbols or digits. In the numeral 3257 the 5 stands for 50; in the numeral 36,289 the 6 stands for 6000.

plus (+) Used to indicate the addition operation, as in $4 + 3 = 7$ (read, "4 plus 3 equals 7").

polygon A closed geometric figure made up of line segments.

prime number A number greater than 1 whose only factors are itself and 1.

product The result of the multiplication operation. In $6 \times 7 = 42$, 42 is the product of 6 and 7.

quadrilateral A four-sided polygon.

quotient The number (other than the remainder) that is the result of the division operation. It may be thought of as a factor in a multiplication equation.

radius (1) Any segment from the center point to a point on the circle. (2) The distance from the center point to any point on the circle.

ray The heavy part of the line shows a ray.

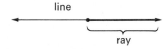

rectangle A quadrilateral that has four right angles.

regrouping A method of handling place value symbols in adding or subtracting numbers.

remainder

Example:
```
     6
  7)47
    42
     5  ← remainder
```

repeated addition Finding the sum of a set of numbers, each of which is the same.

Example: $5 + 5 + 5 + 5$

repeated subtraction Starting with a number and repeatedly subtracting the same given number from each difference that is obtained.

rhombus A parallelogram with 4 sides of the same size.

right angle An angle that has the measure of 90 degrees.

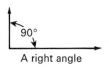

A right angle

right triangle A triangle that has one right angle.

Roman numerals Numerals used by the Romans. Used primarily to record numbers rather than for computing. Examples: IV, IX, XIV.

segment Two points on a line and all the points on that line that are between the two points.

sequence A collection or set of numbers given in a specific order. Such numbers are commonly given according to some rule or pattern.

set A group or collection of objects.

skip count To count by multiples of a given number. Example: Counting by fives — 0, 5, 10, 15, 20, · · · .

solution The number or numbers which result from solving an equation or a given problem.

solve To find the number or numbers which, when substituted for the variable or placeholder, make the given equation true.

square A quadrilateral that has four right angles and four sides that are the same length.

subtraction An operation related to addition as illustrated:

$7 + 8 = 15 \begin{cases} 15 - 8 = 7 \\ 15 - 7 = 8 \end{cases}$

sum A result obtained by adding any set of numbers is referred to as the sum of the numbers.

symmetric figure A figure that can be folded in half so the two halves match.

times (×) Used to indicate the multiplication operation, as in $3 \times 4 = 12$ (read, "3 times 4 equals 12").

triangle A three-sided polygon.

unit An amount or quantity adopted as a standard of measurement.

vertex The point that the two rays of an angle have in common.

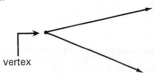

vertex

volume The measure, obtained using an appropriate unit (usually a cube), of the interior region of a space figure.

whole number Any number in the set. {0, 1, 2, 3, 4, 5, 6, 7, 8, 9, 10, 11, 12, 13, 14, · · · }.

zero principle (for addition) Any number added to zero is that same number.

A-43

Tables of Measures

English System — LENGTH	Metric System
12 inches (in.) = 1 foot (ft)	10 centimeters (cm) = 1 decimeter (dm)
3 feet = 1 yard (yd)	10 decimeters = 1 meter (m)
5280 feet = 1 mile (mi)	1000 meters = 1 kilometer (km)
1760 yards = 1 mile	

TIME	
60 seconds (sec) = 1 minute (min)	52 weeks = 1 year (yr)
60 minutes = 1 hour (hr)	12 months (mo) = 1 year
24 hours = 1 day	365 days = 1 year
7 days = 1 week (wk)	366 days = 1 leap year

LIQUID	DRY
8 fluid ounces (oz) = 1 cup	2 pints = 1 quart
2 cups = 1 pint (pt)	8 quarts = 1 peck (pk)
2 pints = 1 quart (qt)	4 pecks = 1 bushel (bu)
4 quarts = 1 gallon (gal)	

English System — WEIGHT	Metric System
16 ounces (oz) = 1 pound (lb)	1000 grams (g) = 1 kilogram (kg)
2000 pounds = 1 ton	1000 kilograms = 1 metric ton

Index

A

Activity Cards, 309-317
Addend(s)
  missing, 56-57, 66
  rearranging, 62-63
  and sums, 54
Addition
  grouping principle for, 60
  multiplication principle, 142-145, 236, 238
  on the number line, 52
  order principle for, 60
  related to subtraction, 50
  repeated, and multiplication, 128-129
  2-digit numbers, 94, 98-99
  see also Addend(s)
Algorithm
  for long division (1-digit divisor), 294
  for multiplication, 242
Angle(s)
  construction of, 80-81
  and parallel lines, 198-199
  right, 80-81
  and triangles, 84-85
Area
  estimating, 19
  square centimeter, 17
  units for measuring, 16-17

C

Center point of a circle, 209
Centimeter
  square, 17
  unit of length, 10
Cents, *see* Money
Chart reading, 108, 111-112, 114
Checking division, 179, 278, 302-303
Coins
  dimes, 92, 229
  dimes and pennies, 33, 90-91, 130-131, 187, 234
  mixed, 112-113
Comparing sets, 58
Coordinate geometry, 256-273
Coordinates, 256
Counting
  and measurement, 4-27
  2-digit numerals, 31
  3-digit numerals, 36-37
  4-digit numerals, 40-41
  5- and 6-digit numerals, 44-47

Cubic units, 25
Cup, 26

D

Diagonals, 206-207
Differences and missing addends, 56-57, 66
Digits, 31, 36-37, 40-41
Dimes, 33, 90-92, 112-113, 131, 229, 234
Distance (miles), 46, 283
Division
  checking, 179, 278, 302-303
  long division algorithm (1-digit divisor), 294
  and multiplication, 179
  and the number line, 176-177, 188, 276-277
  remainders, 300
  and sets, 170-173, 186, 276
  and subtraction, 174-175, 180, 276, 284-289
  see also Quotients
Dollars and cents, 114-115, 293

E

Eighths, 22
Estimating
  area, 19
  products, 250-251
Even
  numbers, 214-216
  numbers, sums and products, 217

F

Factors
  missing, and quotients, 178-180
  and products, 132-133
  rearranging, 140-141
  of ten and one hundred, 226-229
  of ten, twenty, thirty, . . . , 230-231
  two 2-digit, 228-229
  2-digit, 240
  3-digit, 248
  with 4 or more digits, 248-249
  zero through nine (multiplication facts) 146-155, 158, 160
Foot, 8
Fourths, 20-21
Fractions, 14-23, 49, 73, 89, 135, 213, 254, 292
  eighths, 22
  fourths, 20-21

A-45

halves, 18, 20-21
measure to nearest half unit, 14-15
thirds, 22-23
Function graphing, 268-269, 273
Function machine
addition and subtraction, 70-71
coordinate geometry, 268-269
division, 185
even and odd numbers, 215-216
multiplication, 150
multiplication and division, 280
repeated subtraction, 281

G

Gallon, 26
Games, What's My Rule, 68-69
Geometry
angle, 80-81, 84-85, 198-199
circle, 84, 203, 209-210
coordinate, 256-273
diagonal, 206-207
figure, 76-77
graphing negative numbers, 272-273
graphing number pairs, 256-257, 268-269
hexagon, 206-207
hypotenuse, 86
isosceles right triangle, 85
leg, 86
line, 76-77
line segment, 76-79
parallel lines, 196-199
parallelogram, 204-205
pentagon, 206-207
point, 76-77
polygon, 206-207
quadrilateral, 200-203
ray, 76-77
rectangle, 202-203
rhombus, 202-203
right angle, 80-81
right triangle, 84-86
triangle, 82-87
vertex, 74-75
Graph(s), 256-273
Graphing
functions, 268-269, 273
negative numbers, 272-273
number pairs, 256-257, 268-269
Grouping
addends (sums of ten), 63
addends (sums between ten and nineteen), 64-65
principle for addition, 60
principle for multiplication, 140
by tens, 30-31

H

Half inch, measuring to the nearest, 14-15
Halves, 18, 20-21
Hexagon, 206-207
Hundreds (3-digit numerals), 34-37
Hypotenuse, 86

I

Inch, 8; *see also* Measurement
Inequalities (place value), 42-43
Isosceles right triangle, 85

L

Leg of a right triangle, 86
Length, 4-15; *see also* Measurement
Line (line segment), 76-79; *see also* Parallel lines
Liquid measure, 26-27

M

Map, 165, 283
Measurement
of area, 16-21
choosing units of, 8
and counting, 4-27
with different unit segments, 4-11
estimating area, 19
of length, 4-15
of liquid measure, 26-27
to the nearest half inch, 14-15
to the nearest inch, 12-13
units for, 4-27
volume, 24-27
Midpoint of a line segment, 205
Miles, 45, 283
Million, 48
Missing addends and differences, 56-57, 66
Missing factors and quotients, 178-180
Money
cost, 102-103, 235
dollars and cents, 114-115, 293
see also Coins
Multiple(s), 218-219
Multiplication
addition principle, 142-145, 236, 238
algorithm, 242
and division, 179
facts, zero through nine, 146-155, 158, 160
grouping principle for, 140
and the number line, 124-125
one in, 136-137
order principle for, 138

and pairing, 162-163
and repeated addition, 128-129
and sets, 122-123
by ten, 226-229
zero in, 136-137
see also Factors, Multiple(s), Products

N

Negative numbers and graphing, 272-273
Nickels, 112-113
Number line
  addition and subtraction, 52-53
  division, 176-177, 188, 276-277
  multiplication, 124-125
Number theory
  even numbers, 214-216
  factors, 132
  multiples, 218-219
  odd numbers, 214-216
  prime numbers, 222-223
Number(s)
  even and odd, 214-216
  negative, 272-273
  pairs and graphing, 256-257, 268-269
  prime, 222-223
  sets and numerals, 30-31
Numerals
  sets, numbers and, 30-31
  2-digit numerals and counting, 31
  3-digit numerals, 36-37
  4-digit numerals, 40-41
  5- and 6-digit numerals, 44-47

O

Odd
  numbers, 214-216
  numbers, sums and products, 217
Odometer, 46
Order principle
  for addition, 60
  for multiplication, 138
Ounce (liquid), 26-27

P

Pairing and multiplication, 162-163
Pairs, number and graphing, 256-257, 268-269
Parallel
  lines, 196-199
  lines and angles, 198-199
Parallelogram, 204-205
Pennies, 33, 90-91, 130, 187, 234
Pentagon, 206-207

Pint, 26-27
Place value
  counting, 4-27
  grouping by tens, 230-231
  hundreds, 36-37
  inequalities, 42-43
  million, 45
  odometer, 46
  pennies and dimes, 33, 90-91, 130, 187, 234
  thousands, 38-41, 44-47
  2-digit numerals, 31
Point, 76-77
Polygons, 206-207
Prime numbers, 222-223
Principle(s)
  grouping principle for addition, 60
  grouping principle for multiplication, 140
  one principle for multiplication, 136-137
  multiplication-addition, 142-145, 236, 238
  order principle for addition, 60
  order principle for multiplication, 138
  zero principle for multiplication, 136-137
Product sets, 162-165
Products
  estimating, 250-251
  and factors, 132-133
  related, 230-231
  and sums, even or odd, 217

Q

Quadrilaterals, 200-203
Quart, 26-27
Quarters, 112
Quotients
  finding, 171-173
  and missing factors, 178-180
  subtracting to find, 174-175, 180, 276, 284-289
  with zero endings, 278-279

R

Ray, 76-77
Rearranging
  addends, 62-63
  factors, 140-141
Rectangles, 202-203
Remainders in division, 300
Rhombus, 202-203
Right angles, constructing, 80-81
Right triangles, isosceles, 85
Rule
  What's My Rule games, 68-69
  see also Function
Ruler, centimeter, 10-11

S

Segment
  line, 76-79
  for measuring, 4-11
Sets
  comparing, 58
  division, 170-173, 186, 276
  multiplication, 122-123
  numbers and numerals, 30-31
Simple closed curve, 208-209
Square
  centimeter, 17
  unit, 16-17
Subtraction
  and division, 174-175, 180, 276, 284-289
  on the number line, 52-53
  relation to addition, 50
  2-digit numbers, 66-67, 94-95, 106-107
Sums
  and addends, 54
  and products, even or odd, 217
  between ten and nineteen, 63-65
  of ten, twenty, thirty, forty, . . . , 92-93
  of 2-digit numbers, 92-95, 98-99
Symmetry, 210-211, 264-265

T

Tens
  finding sums of, 63
  grouping by, 30-31
  sums of ten, twenty, thirty, forty, . . . , 92-93
Thirds, 22

Thousands (5- and 6-digit numerals), 38-41, 44-47
Time, 245
Triangle
  angles and, 84-85
  definition of, 82
  right, 84-87
  right isosceles, 85

U

Units
  choosing, 8
  for measuring, 4-11
  see also specific units

V

Vertex, 74-75
Volume
  by counting cubic units, 24-25
  liquid measure, 26-27
  units of, 24-27
  see also Measurement

Y

Yard, 8

Z

Zero
  endings, quotients with, 278-279
  in multiplication, 136-137
  negative numbers and graphing, 272-273